CHANGING MIND

Transitions in Natural and Artificial Environments

STUDIES OF NONLINEAR PHENOMENA IN LIFE SCIENCES

Editor-in-Charge: Bruce J. West

Vol. 1: Fractal Physiology and Chaos in Medicine
by B J West

Vol. 2: Patterns, Information and Chaos in Neuronal Systems
edited by B J West

Vol. 3: The Lure of Modern Science — Fractal Thinking
by B J West & B Deering

Vol. 4: Physical Theory in Biology — Foundations and Explorations
edited by C J Lumsden, W A Brandts & L E H Trainor

Vol. 5: Nonlinear Dynamics in Human Behaviour
edited by W Sulis & A Combs

Vol. 6: The Complex Matters of the Mind
edited by F Orsucci

Vol. 7: Physiology, Promiscuity, and Prophecy at the Millennium: A Tale of Tails
by B J West

Vol. 8: Dynamics, Synergetics, Autonomous Agents
edited by W Tschacher & J-P Dauwalder

Studies of Nonlinear Phenomena in Life Science – Vol.9

CHANGING MIND

Transitions in Natural and Artificial Environments

Franco F. Orsucci

Institute for Complexity Studies, Rome, Italy

World Scientific

New Jersey • London • Singapore • Hong Kong

Published by

World Scientific Publishing Co. Pte. Ltd.

P O Box 128, Farrer Road, Singapore 912805

USA office: Suite 1B, 1060 Main Street, River Edge, NJ 07661

UK office: 57 Shelton Street, Covent Garden, London WC2H 9HE

British Library Cataloguing-in-Publication Data
A catalogue record for this book is available from the British Library.

CHANGING MIND
Transitions in Natural and Artificial Environments

ISBN 981-238-027-2

This book is printed on acid-free paper.

Printed in Singapore by Mainland Press

Preface

The elephant in the dark.

This is the Sufi tale, narrated by Rumi, the poet:

"*The Elephant in the dark, on the reconciliation of contrarieties*

Some Hindus had brought an elephant for exhibition and placed it in a dark house. Crowds of people were going into that dark place to see the unknown beast. Finding that ocular inspection was impossible, each visitor felt it with his palm in the darkness.

The palm of one fell on the trunk.
'This creature is like a water-spout,' he said.
The hand of another lighted on the elephant's ear.
To him the beast was evidently like a fan.
Another rubbed against its leg.
'I found the elephant's shape is like a pillar', he said.
Another laid his hand on its back.
'Certainly this elephant was like a throne', he said.

The sensual eye is just like the palm of the hand. The palm has not the means of covering the whole of the best. The eye of the Sea is one thing and the foam another. Let the foam go, and gaze with the eye of the Sea. Day and night foam-flecks are flung from the sea: amazing! You behold the foam but not the Sea.
We are like boats dashing together; our eyes are darkened, yet we are in clear water" (Rumi, 1995).

Rediscovering mind

We have discovered mind since prehistoric shamans, their graffiti in caves and Lucy's first insights in African savannas. We always discover again its amazing qualities and failures every morning, when we wake up. But we still don't really know how mind functions; if it is in the skull, everywhere or elsewhere in the body.

A few years ago John Horgan (1999) wrote a book that has had interesting titles in its different versions: *The undiscovered mind. How the human brain defies replication, medication and explanation* (US version); *How the brain defies explanation* (UK version); *The inviolate mind. A challenge for psychology and neurosciences* (Italian version). Probably there are many other versions: this book is about the fragmentation of studies on mind, and their primitive state of development.

These titles are symptoms of a persistent difficulty in considering the state of the art of mind sciences as a set of harmonic and cooperative studies. They are still divided in a myriad of disciplines and sub disciplines, often in conflict and using different technical languages. This, obviously, originates problems in sharing or translating knowledge between scholars all involved in the same challenge for knowledge.

The project of cognitive science, in its endeavour to interconnect seemingly disparate disciplines such as linguistics, anthropology, artificial intelligence, neuroscience, psychology and philosophy, has been representing the challenge in unifying all the studies on the nature of subjectivity and the research on mind structures (Gardner, 1985). It generated ambitious questions such as: *what is knowledge*?

The enterprise is crucial but not new. Classical philosophy was about this until the *Naturwissenschaften* (natural philosophy), before Cartesian and neo-Cartesian positivisms conquered the control of operations.

The Sufi tale is clear: when the object is so large and complex, and you are not able to encompass it, your perspectives can be partial and misleading. But, if you try to deal with the whole big object you should be sure that you wouldn't be devoured or crushed.

Applications

This has been even truer in the field of applicative cognitive and neuro-science. Clinical psychology, psychiatry, neurology and the human sciences were fragmenting in a myriad of small schools. The emphasis of each one in defending its vantage point was proportional to its partiality (i.e. ideology). The reason was probably embedded in the contamination of the authentic scientific motivation with ideological and economical considerations due to partiality of results, and the lack of methodological rigour in formalization and empirical verifications.

This has been related to the lack of the appropriate mathematical and statistical tools to handle such a big beast.

Dynamical systems theory

The gradual emergence of a set of formal and methodological tools called Dynamical Systems Theory, or Complexity Theory could finally make the scene different. This is called also Nonlinear Science: a marker of the shift in scientific paradigms (Kuhn, 1996) because all dynamical systems that are not linear are called nonlinear.

This name by exclusion might seem surprising. As Stanislaw Ulam remarked: "Calling a science 'nonlinear' is like calling zoology 'the study of non-human animals.' " This means that almost every system is complex, dynamical and nonlinear by nature. It is (the limit of) our approaches that make us see them as linear. Yet the distinction has been necessary in the natural sciences, which became so accustomed to linear systems, because they were more *treatable*. The discovery of the possible scientific study of dynamical behaviors unrestricted by linearity is one of the greatest scientific revolutions of all times. It is becoming even a revolution in our everyday perception of reality, as trees and lightning become "scientific objects" just as cubes and cones and even dodecahedrons.

A state of the art

In fact, the science of nonlinearity still resembles a sea of ignorance with some small islands where results are known and applicable. The important thing is that now we know that this sea exists and we can explore it. This is well represented in the sketch below that we owe to Thomas Schreiber (Kantz & Schreiber, 1997) and Francisco Varela (1991).

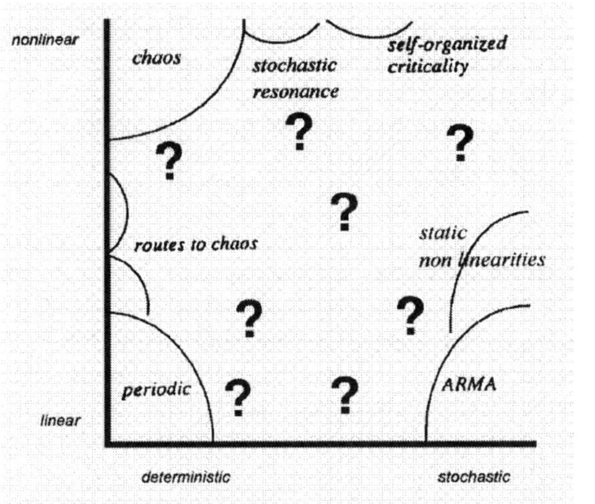

A state of the art in complexity theory (Varela, 1991; Kantz, 1997)

These small islands of knowledge called *chaos, SOC or stochastic resonance* are like candles in a total darkness: we can finally have intuitions of the elephant's whole shape. Or at least knowing that we are dealing with that big elephant, and not with some bizarre furniture.

Theory

The book includes some maps deriving from a personal route through cognitive science: philosophy, medicine, psychiatry, linguistics, psychoanalysis, and nonlinear dynamics: past and current studies.

The land to discover, in a collective project, is *a general theory of mind and its applications (i.e. change)*:

1) mind multilayered structures;
2) mind couplings;
3) modelling;
4) transitions in natural and artificial environments.

Bonds

The object of knowledge is where there are no objects: in the bridges of relationships. This means tracing the sketch of *a biophysical theory of human relations*. As we know that "perfection is the enemy of excellence", the difficulty of the enterprise will be an additional source of passion for further explorations.

The following is another sketch that the painter, architect and sculptor Gianlorenzo Bernini made for a *Pietà,* in a quasi topological and dynamical presentation of a coevolving interaction. It confirms that in a complicated field, as mind sciences obviously are, artists can often reach the metaphorical knowledge that scientists are trying to grasp in other formal ways (Verhulst, 1994). Art and metaphors sometimes will help us, because their icons have the power to encompass, in a single form, huge amounts of data.

Sketch for a sculpture of a *Pietà* by Gianlorenzo Bernini, intuitive presentation of a coevolving interaction

Acknowledgements

I acknowledge with gratitude many useful exchanges of ideas on various aspects related to this book with:
Fred Abraham, Diederick Aerts, Daniel Amit, Uwe an der Heiden, Tito Arecchi, Dick Bird, Stefano Boccaletti, Pietro Bria, Francesco Corrao, Fabrizio Davide, Sergio De Risio, Walter Freeman, Liane Gabora, Alessandro Giuliani, Jeffrey Goldstein, Steve Guastello, Matthjis Koopmans, Juergen Kurths, Gry Kristin Lia, Sebastiano Maffettone, Terry Marks-Tarlow, Tullio Minelli, Susan Mirow, Giovanna Morgavi, Laura Olivetti, Lydia Pallier, Giorgio Parisi, Lou Pecora, Luciano Pietronero, Robert Porter, Lorena Preta, Sergio Rinaldi, Robin Robertson, Giuseppe Ruggeri, Guenther Schiepeck, Gary Bruno Schmidt, Thomas Schreiber, Harold Searles, Bill Sulis, Graeme Taylor, Roberto Tagliacozzo, Karl Toifl, Irina Trofimova, Nicoletta Sala, Franco Scalzone, Alan Stein, Francisco Varela, Angelo Vulpiani, Gemma Zontini.
I am grateful to Kok Khoo Phua and Bruce West for hosting this book in the Studies of Nonlinear Phenomena in Life Science series at WSP.
Finally, I am grateful for the perceptive and collaborative attention of Lakshmi Narayan, Yolande Koh and Kim Tan of World Scientific Publishing.

References

Gardner, H. (1985). *The mind's new science, a history of the cognitive revolution*. New York: Basic Books.

Horgan, J. (1999). *The undiscovered mind, how the human brain defies replication, medication, and explanation*. New York: Free Press.

Kantz, H. & Schreiber, T. (1997). *Nonlinear time series analysis*. Cambridge: Cambridge University Press.

Kuhn, T. S. (1996). *The structure of scientific revolutions*. Chicago, IL: University of Chicago Press.

Rumi, J. J.-D. & Barks, C. (1995). *The essential Rumi*. San Francisco, CA: Harper.

Varela, F. J., Thompson, E., & Rosch, E. (1991). *The embodied mind, cognitive science and human experience*. Cambridge, Mass: MIT Press.

Verhulst F (1994) Metaphors for psychoanalysis. *Nonlinear Science Today* 4 (1).

Contents

Chapter 1: E-Mergence

The Golden Chain:
Zeus was anxious to know
How shall all things be one,
yet each distinct?
Callimachus

1.1 E-Mergence

Following some instructions that Ludwig Wittgenstein (1967) has offered about knowledge games, we try first to draw the lines in the field and recall the rules of the game we are going to play. Most of the game will be played in the field defined by two conceptual basic poles in system dynamics: *emergence* and merging, or *fusion*. While the first one is more frequently cited and discussed, the second one is often neglected. It seems quite strange, because it is so difficult to define and understand the first without the second and vice-versa. We can say more: they form a dynamic couple. If we start from their etymologies and semantic fields in natural language, we can easily see why (Merriam-Webster, 1998).

Merge – etymology: Latin *mergere*; akin to Sanskrit *majjati* ('he dives').

Transitive senses:

- Archaic, to plunge or engulf in something, immerse;
- to cause to combine, unite, or coalesce;
- to blend gradually by stages that blur distinctions;

Intransitive senses:

- to become combined into one;
- to blend or come together without abrupt change;

Mergence \ noun

Synonyms: mix, amalgamate, blend, compound, fuse, interblend, interfuse, intermingle, meld, mingle.

Emerge – etymology: Latin *emergere*, from *ex* + *mergere* to plunge.

Senses (intransitive):

- to become manifest
- to rise from or as if from an enveloping fluid out into view
- to rise from an obscure or inferior position or condition
- to come into being through evolution

Synonyms: appear, loom, show.

Related Words: derive, originate, spring, stem; arise, materialize, rise; come (forth), come out, emanate, flow, issue (forth); proceed.

Idioms: appear on the horizon, come on the scene, come out in the open, come to light, make its appearance.

Contrasted Words: disappear, fade, fade (out); evaporate; dissolve.

From a first comparison we find an interesting thing, it seems that emerge is a semantic evolution (see Sanskrit and Latin) from merge.

We could say that the dialectics

$$M \Leftrightarrow E$$

represents a dynamic relation between some whole and its part(s).

For example, following the Sanskrit original etymology, if I dive in the sea I become part of it, for an observer, but during my swimming I emerge and I can be differentiated from it. And, also, I am not caused by the sea; I just have a dynamic relation with it. The simplicity or complexity of an object and its consistency depends from the perspective and resolution the object is considered.

Ernest Nagel (Nagel, 1961) cites Mill's System of Logic (1843) as the *locus classicus* on the notion of emergence. For Mill, the key to the distinction between emergent and non-emergent properties centers on a distinction regarding two different ways in which conjoint causes can produce an effect: Non-emergent properties are effects that are mere summations of the effects of each of the causal conjuncts, whereas emergent properties are effects that are not sums of the effects of each causal conjunct.

Following (Lalande, 1962) the term emergence was used mainly to designate the fact that one thing comes from another *without a simple relation of cause and effect between the twos: in these terms it would imply self-organization and/or a different activity of the observer*.

Lloyd Morgan (1922) wrote: “The emergence of a new quality from any level of existence means that, at that level, there comes into being a certain *constellation* or collocation of the motions belonging to that level, and possessing the quality appropriate to it; and this collocation possesses *a new quality distinctive of the higher complex*. The quality and the constellation to which it belongs are at once new, and expressible without residue in terms of *the process proper to the level from which they emerge; just as mind is a new quality distinct from life, with its own peculiar methods of behaviour...not merely vital, but also vital."* (Morgan, 1923) (Italics are mine).

1.1.1 The emerging mind

A system exhibits emergent properties when those properties are more than the sum of its parts' properties. We could posit that mental properties emerge from certain complex sets of physical properties, for instance, physical properties of human brains. Thus mental events are not identical to any brain events but instead emerge from them. *After the current studies on neural plasticity we can also say that brain emerges from mind, because it is moulded by mental events. As in the Moebius band, container and content in the mind-brain relationship can be reversed in their reciprocal function depending from the perspective of the observer.* That is, an increase of knowledge about the way certain effects are obtained may reveal that certain effects are decomposable into the effects contributed by subcomponents of that system.

We have to consider both the whole (decomposable), its parts (fusible), an observer function. There is also a possibility that the observer is observed by the whole, in this way becoming a whole object, and/or mirrored.

We may recall the book written by the Sufi Persian mystic Farid al-Din Attar, called "The Conference of the Birds", which inspired also a theatre play by Peter Brook. This story is about thousands of birds that set out to find a King for their Kingdom. They cross 7 valleys, and each valley has a different name and significance. Some birds give up, some die, some decide to return, and many make excuses. Finally, 30 birds make it and when they reach the end they realize that the King or Simurgh they came to look for is merely a mirror. In Farsi (the Persian language), *si* means 30 and *murgh* means bird, so... Attar the author, very cleverly named the King "Simurgh". Therefore, 30 birds meet 30 birds and realize that we are all part of a whole... so, God is in us and we are in God, we are not separate from God. This is a very basic Sufi belief. As a word of explanation, the Sufi ideal is to worship the divine by loving it. Thus the devotee's relation to God is depicted as that of lover and beloved. It is a consuming and transformative love; an embracing so intense that all distinctions of *self* and *other* disappear before it. (……)

In the study of complex systems, the idea of emergence is used to indicate the arising of patterns, structures, or properties that do not seem adequately explained by referring only to the system's pre-existing components and their interaction. Emergence becomes of increasing importance as an explanatory construct when the system is characterized by the following features:

- when the organization of the system, i.e., its global order, appears to be more salient and of a different kind than the components alone;
- when the components can be replaced without an accompanying decommissioning of the whole system;
- when the new global patterns or properties are radically novel with respect to the pre-existing components;
- thus, the emergent patterns seem to be unpredictable and non-deducible from the components as well as irreducible to those components.

1.1.2 Pre-Socratic shamans

Since the Pre-Socratics, the basic problem of natural philosophy has been to discover how order and form arise from turbulent and disordered states of mind and matter (as we will see, the distinction between the twos for them is often fuzzy).

It is the problem of *origin* and *primary causes of order and organization: how did it originate, what were its causes, and what does it really mean?*

The start of these studies is immersed in the shining mist of prehistoric Shamans' culture, and their followers in Asia and Europe. In classical times shamanistic practice and ideology extended across the steppes into the northern territories of the Indo-European tribes, from China and India to Persia and Greece (Cornford, 1957). The history of Chinese and Indian philosophy, growing from the shamanistic-animistic cultures, goes in parallel with the Western paths. The same roots are the core of the *Urgeist*, the primal spirit of creation in civilizations.

Mircea Eliade (1968), in his classic study on shamanism, includes between the shaman powers the following masteries: flying, going underground, power dance, ecstasies, revelations, talking with animals, managing weather, healing, and knowledge of plants. These powers are metaphorically duplicated in their knowledge of the world. Shamanism became a recognized public institution in Western and Eastern societies and produced mythical and metaphysical theories concerning creations and the relationships between human psychobiological beings and the natural realm.

These constructions of knowledge in the "primal stuff", the *primal mind-body-world matrix,* were often motivated by applicative purposes, i.e. helping the social group to master primary needs like food, war or diseases. Disease theories are very interesting because they are the result of a difficult cognitive task: explaining a (disturbing) change in the primal mind-body-world matrix. The appearance of a disturbance causes the need to explain and to differentiate, to draw boundaries in this primal quasi-continuum. If we take a look at some archaic theories of disease[7] and the related shamanistic therapies, we find that the therapies were intended to recover some order and proper balance in the continuum of the primal matrix. Ellenberger (1970)

DISEASE THEORY	**PERIOD**	**GEOGRAPHY**	**THERAPY**
Object intrusion	Palaeolithic	Old world	Extraction of disease object
Loss of soul	Late Palaeolithic	Siberia	Find and restore soul
Spirit intrusion	End Pleistocene	Western Asia	Exorcism, extraction, transference
Taboo violation	Recent	Three areas	Confession, propitiation
Sorcery	Recent	Three areas	Counter magic

Modified table from Ellenberger, 1970.

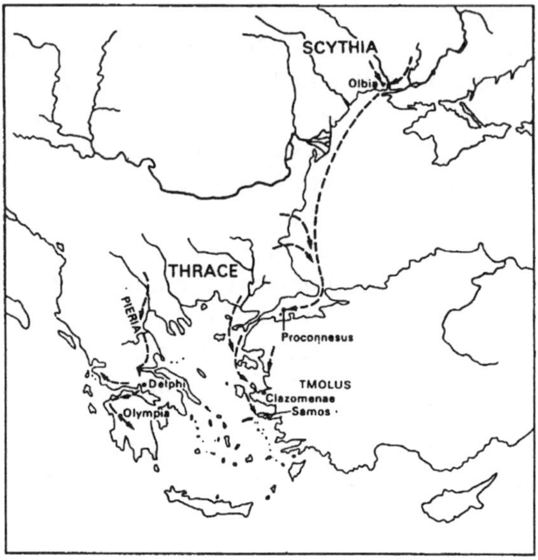

Patterns of shamanistic influence during Bronze Age in Archaic Greece
(West ML, 1983)

The map (West ML, 1983) shows the shaman routes along which the Asian shamanistic culture was in current contact with the Greek culture at the beginning of its "historic" evolution and after. Even in classical times shamanistic practice and ideology extended across the steppes into the northern territories of Indo-European tribes. While in China the shamanistic influence was reflected in Taoism and in Japan in Shintoism, in Greece traces bring to the Orphic-Dionysian cult. In the Ionian region we find the main archaic Greek shamans: Aristeas of Proconnesus, Hermotimus of Clazomenae, and Pythagora of Samos who claimed to be the Hyperborean Apollo. In this region we locate the development, in the sixth century of a Dionysian cult which adopted Orpheus as its prophet. The last trail leads from Ionia to Sicily and Italy where Pythagoras was bringing the Orphic rites; Parmenides used shamanistic imagery in his philosophical poems; and Empedocles offered prophecies, cures for diseases, control of wind and rain, and the ability to raise the dead.

Friedrich Nietzsche (1873) condensate this morphogenetic (metaphoric) research:

"In reality, however, that hero is the suffering Dionysus of the mysteries, that god who experiences the suffering of the individual in himself, the god about whom the amazing myths tell how he, as a child, was dismembered by the Titans and now in this condition is venerated as Zagreus. Through this is revealed the idea that this dismemberment, the essentially Dionysian suffering, is like a transformation into air, water, earth, and fire, that we also have to look upon the condition of individuation as the source and basis for all suffering, as something in itself reprehensible. From the laughing of this Dionysus arose the Olympian gods, from his tears arose mankind. In that existence as dismembered god Dionysus has the dual nature of a cruelly savage daemon and a lenient, gentle master."

Theogonies are between the first steps in this evolution: they are poems which give an account about the origin of the world and the history of gods and heroes. The best-known Greek Theogony is that of Hesiod. Anyone at all familiar with Greek mythology will recognize Hesiod's genealogy. The poems of Hesiod, in the Western civilization, are one of the first artistic representations of these transitions, and they portrait the crucial role of mythological and metaphorical narrations in this area of human knowledge.

The Orphic Theogonies differ from Hesiod's, and can appear somewhat bizarre to a modern reader. There seem to have been several Orphic Theogonies, but most have not survived. The Rhapsodic Theogony does survive in fragments. It is a composite work based on earlier Orphic poems; fragments of various related or ancestral works have been discovered which date back to the classical period, three or four hundred BC.

1.1.3 Birth of Western natural philosophy

Other examples of dealing with morphogenesis problems in the first steps of *Western natural philosophy* are given by the so-called Pre-Socratic (Diehls, 1966; Mainzer, 1997). We find the first seeds of a new order growing the *mind-body-world primal matrix*. As if they were exploring a new and wild land they draw new and arbitrary boundaries on wild landscapes they were occupying. A schizophrenic used to explain his origins as a dynamical process in which he was springing from, and sometimes re-diving in a ball of clay. This ball was inhabited by primitive small men, and sometimes he was one of them.

The same focus on the origins of the world, drawing boundaries on virtual landscapes, is present also in psychotic delusions. In this case we find a regression to the archaic origins of thinking.

Thales of Miletus (625-545 B.C.), the same who is said to have proven the first geometric theorems, is also the first to believe that only material causes can be at the origin of everything. No genuine fragment survives from his writings, but we can say that Thales assumes that water, or *the wet*, is the first cause of everything. His argument is the observation that nourishment and the seeds of all being are wet, and the natural substratum of all wet things is water (we may suppose some relation with the social evolution from hunting to agriculture).

Anaximander (610-545 B.C.), his student and companion, extends and complexifies Thales assumptions: the primary core is not just in one substance, but matter originates in the continuous tensions of opposites. Heat vs. cold; dry vs. wet…he assumes that the first cause is a "*boundlessly indeterminable"* matter without form, the *àpeiron*, out of which the forms of matter emerge.

"The everlasting generative matter split apart in the creation of our world and out of it a sphere of flame grew around in the air surrounding the earth like the bark around a tree; then, when it tore apart and bunched up into definite circles, the sun, moon and stars took its place.

It is a cosmogony, and it includes even a biological evolution, because he assumes that first humans were rising from the sea (idea he derived from observations he did on sharks). "The Non-Limited is the original material of existing things; further, the source from which existing things derive their existence is also that to which they return at their destruction, according to necessity; for they give justice and make reparation to one another for their injustice, according to the arrangement of Time" (Diehls, 1949).

The third famous Milesian philosopher is Anaximenes (…-525 B.C) who regards *change* as the effect of the external forces of condensation and rarefaction: every kind of matter can serve as basic. But, he prefers *air*: "And rarefied, it became fire; condensed, wind; then cloud; further, by still stronger condensation, water; then earth; then stones; but everything else originated by these".

Heraclitus of Hephesus (ca. 500 B.C.), "*the obscure*" as he was often called, took over from Anaximander the doctrine of struggle of opposites, writing a book on every known doctrine, "in oracular style". The original matter is itself changing, identified with fire:

"The ray of lightning guides the All. This world order which is the same for all was created neither by one of the gods nor by one of the humans, but it was always, and it is, and will be eternally living fire, glimmering and extinguishing according to measures".

As Heisenberg (1952) declared, the Heraclitus' pronouncements can be considered extraordinarily modern: we just have to substitute the word "fire" with "energy" and the result is modern physics. And the opposites are held in unity by hidden harmony (*armonìa*), the cosmic law which he called *logos*.

If the struggle comes to an end, according to Heraclitus, the world comes to a final state of equilibrium, and Parmenides of Elea (ca. 500 B.C.) starts from there. He described this state of matter in which there are no changes and motions as the primal state. Matter is distributed homogeneously and without any preferred motion (isotropically). According to Umberto Eco's personal experience, one of his students described this view of the world in this preoccupied way: "Parmenides asserts there only one thing…and I am not part of it!"

The Eleatic school considered the Heraclitus prominence given to change and motion as an illusion of senses, but the two views are not necessarily alternative.

After water, air and fire Empedocles of Syracuse (492-430 B.C.) added *earth* as the fourth basic element. These elements are free to mix and bind, to dissolve and separate. But according Empedocles there were two dynamic principles to determine their movements: nature and chance (*tùke*). Mixing and separation, merging and emerging between these four natural elements were causing changes in natural forms.

It is again amazing the resonance between these explanations of the natural world between Western and Eastern philosophies. Anaxagoras (499-426 B.C.) maintained a mixed theory of matter, but replaced the four Empedocles' elements with an unlimited number of substances, composed of seed particles (*spèrmata*). They were granulated (digital) but infinitely divisible, and in this way he prepared the way for the atomist theories.

1.1.4 Maps

Leucippus of Abdera was credited with one book called *The Great World-Order* (later attributed to Democritus) and perhaps a monograph *On Mind*.

Certain terms of the Atomic Theory probably can be traced back to Leucippus, e.g.

- Atoms (uncuttables)
- Close pressed (units)
- Great Void
- Absection
- Rhythm (*meaning form)*
- Contact (*meaning arrangement*)
- Aspect (*meaning position*)
- Entanglement

(Diels, 1948)

Leucippus was the founder of Atomism. We know next to nothing about his life, and his book appears to have been incorporated in the collected works of Democritus. No writer subsequent to Theophrastus seems to have been able to distinguish his teaching from that of his more famous disciple. Indeed his very existence has been denied, though on insufficient grounds. Aristotle gives a clear and intelligible account of the way Leucippus' theory arose. It originated from Parmenides' denial of the void, from which the impossibility of multiplicity and motion had been deduced.

Leucippus supposed himself to have discovered a theory which would avoid this consequence. He admitted that there could be *no motion if there was no void*, and he inferred that it was wrong to identify the void with the non-existent. Leucippus was the first philosopher to affirm, with a full consciousness of what he was doing, the existence of empty space. The Pythagorean void had been more or less identified with 'air', but the void of Leucippus was really a vacuum. Besides space there was body, and to this Leucippus ascribed all the characteristics of Parmenides notion of the real.

The assumption of empty space, however, made it possible to affirm that there was an infinite number of such reels, invisible because of their smallness, but each possessing all the marks of the Parmenidean One, and in particular each indivisible like it. These moved in the empty space, and their combinations can give rise to the things we perceive with the senses. Pluralism was at least stated in a logical and coherent way.

Democritus was born at Abdera, about 460 BC, although according to some 490. His father was from a noble family and of great wealth, and contributed largely towards the entertainment of the army of Xerxes on his return to Asia. As a reward for this service the Persian monarch gave and other Abderites presents and left among them several Magi. Democritus, according to Diogenes Laertius, was instructed by these Magi in astronomy and theology. Petronius relates that he was acquainted with the virtues of herbs, plants, and stones, and that he spent his life in making experiments upon natural bodies. He acquired fame with his knowledge of natural phenomena, and predicted changes in the weather. He used this ability to make people believe that he could predict future events.

Democritus has been commonly known as "The Laughing Philosopher", and it is gravely related by Seneca that he never appeared in public with out expressing his contempt of human follies while laughing. Democritus compared the motions of the atoms of the soul to that of the particles in the sunbeam which dart hither and thither in all directions even when there is no wind, and we may fairly assume that he regarded the original motion of the other atoms in much the same way.

The atoms are not mathematically indivisible like the Pythagorean monads, but they are physically indivisible because there is no empty space in them. Theoretically, then, there is no reason why an atom should not be as large as a world. Such an atom would be much the same thing as the Sphere of Parmenides, were it not for the empty space outside it and the plurality of worlds. As a matter of fact, however, all atoms are invisible. Our first steps in Western philosophy of nature, which is defining different ways to map order and disorder in what we have called the primal matrix, justify a brief contemporary digression on ways to define parameters for *order and control.*

We might remind that, the global variables, which "order" the smaller components in ways that accommodate the demands of energy optimisation, are call *order parameters*. It characterizes the order of a phase of a system, for example its transition temperature. An *order parameter* has a nonzero value below the transition temperature and a zero value above it. If the phase transition is continuous the order parameter goes to zero continuously as the transition temperature is approached from below. Disorder parameters are nonzero above the transition, and zero beneath it. Order parameters are associated with the *(broken) symmetry* of a system. A situation in which the ground state of a many body system or vacuum state of a relativistic quantum field theory has a lower symmetry than the Hamiltonian or Lagrangian defining the system. Order parameters define structure and boundaries of a given system.

The variables that capture the accommodating properties and relations of the components within a given state are called *control parameters*.

Thus, *order parameters provide an overall specification of different system phases, whereas control parameters represent variables that are dynamically adjusted to provide structure and stability within phases.*

1.1.5 *Klinàmen*

Epicurus was born around 341 BCE, seven years after Plato's death, and grew up in the Athenian colony of Samos, an island in the Mediterranean Sea. He was about 19 when Aristotle died, and he studied philosophy under followers of Democritus and Plato. Epicurus founded his first philosophical schools in Mytilene and Lampsacus, before moving to Athens around 306 BCE. There Epicurus founded the Garden, a combination of philosophical community and school where the residents put Epicurus' teachings into practice.

The modification of Democritus' views proposed by Epicurus is the addition of the 'swerve' (*klinàmen*) to atomic movement. In addition to the regular tendency of atoms to move downward, Epicurus thinks that occasionally, and at random times, the atoms swerve to the side. One reason for this swerve is that it is needed to explain why there are atomic collisions. The natural tendency of atoms is to fall straight downward, at uniform velocity. If this were the only natural atomic motion, the atoms never would have collided with one another, forming macroscopic bodies.

As Lucretius, the most important Roman follower of Epicurus, puts it, they would fall downward, like drops of rain, through the deep void.' So, Lucretius suggests that sometimes, in uncertain times and places, a minimal deviation, the *klinàmen*, perturbs the eternal and universal fall of atoms. This will generate a vortex, origin of a new world and of natural entities. The second reason for thinking that atoms swerve is that a random atomic motion is needed to preserve human freedom and 'break the bonds of fate,' as Lucretius says.

If the laws of atomic motion are deterministic, then the past positions of the atoms in the universe, plus these laws, determine everything that will occur, including human action. Cicero reports that Epicurus worries that, if it has been true from eternity that, e.g., "Milo will wrestle tomorrow", then presently deliberating about whether to make it true or false would be idle.

The configuration of atoms has been compared with the sequence of letters in words which has led to the presumption that atomistic ideas could be developed only in cultures using phonetic alphabets. These alphabets are strongly digital in comparison the iconic structures (Serres, 1995) of pictographic or ideographic alphabets. In fact, in China, where there were just ideographic characters, the particle idea was unknown and a field-and-wave conception of the natural processes prevailed.

Serres sees in Lucretius' text a premonition of 'post-modern sciences;' notably chaos theory - of which he considers De Rerum Natura a "first example" and in whose reference he reads the *klinàmen* - and the theory of complexity. Serres builds on *klinàmen* (in a reference back to Lacan) a theory of trauma that rests on the concept of chance (Aristotle's' differentiation between necessity on the one side and *tyke* and *automaton* on the other). The infinitesimally small and contingent time of the *klinàmen* designates the time of the trauma core, even while it heralds the birth of complexity: the birth of the world from the spirit of trauma, and the necessary trauma of birth.

1.1.6 Symmetry

Plato, in his dialogue *Timaeus*, introduced the first mathematical model of atomism. *Thaetetus* made a complete determination of all the regular bodies that are possible in a 3 dimensional space: thus representing the Empedocles world in geometrical terms. Then Plato developed a kind of elementary particle physics in which elements are transformed into each other according to geometrical laws. We have here the first attempt in the history of natural science to explain matter and its states by simple geometrical laws. As (Schrödinger, 1967) observed: "...the elementary particles have the form Plato ascribed to them because it is the mathematically most beautiful and simplest form. Therefore the ultimate root of phenomena is not matter, but instead mathematical law, symmetry, mathematical form..."

In Antiquity and the Middle Ages Plato's mathematical atomism gained little support, mainly because of some *aporia* showed in the *Timaeus* itself. In fact, how are the functions and forms of living organisms to be explained? The theory of regular solids it seems formed by rigid and dead forms in comparison with the ever pulsating and transforming living things. There were also questions about relations between form and function to be explained. This is where the Aristotelian philosophy of nature begins.

Aristotle proposed equilibrium in nature mainly on the basis of the ways in which living organisms such as plants and animals function. Aristotle called the general principle "form" (*èidos*) which was shaping "matter" (*ùle*). These were the basic principles of nature derived by inductive abstraction. Matter was characterized by the potential force for being formed (*dùnamis*) and produce reality (*energèia*).

1.1.7 *Physiologòi*

Diogenes Laertius divides the ancient philosophers in two groups, Ionian and Italiote, headed by Anaximander and Pythagoras (Diels, 1948). In the Ionian schools some form of science comes to its climax in Atomism: Gods disappear; Soul is reduced to a dust of material particles; Life has gone out of Nature. This is the fate of a science inspired by Euclidian geometry, and neglectful of biology. This is to affirm the idea that science is a simplifying device to exclude the subject of knowledge from the rational world-picture to be constructed. This attitude originated from the Ionian *Physiologòi*. The exclusion of the subject became an inherent habit of the making of science in the next centuries of Western thinking. They affirm the principle that a (scientific) discovery must be independent from the author's subjectivity, in so being fully objective. We could even say that in this "positive" way to produce knowledge, the best theory or discovery has no author (or the author is unknown), because it pre-existed in the objective world. This is a way to reach a full accomplished separation between subject (mind) and world.

Behind Anaximander stands Homer with his troop of luminous Olympians; behind Pythagoras we discern the troubled shapes of Orpheus and Dionysus: two different attitudes towards the life of man and nature. A thread of continuity can be traced back from the final achievement of deterministic science, the representation of a world of individual atoms governed by *Necessity and Chance*, to the world of Olympian Gods subordinate to *Destiny, Moira.*

Philosophical writings of the sixth and fifth centuries were commonly entitled *Concerning Nature, Perì Physeos.* 'Nature' is the name they gave to the ultimate stuff, from which the world of things has arisen and will perish again. We have no satisfactory rendering for *physis*: primary substance is charged with Aristotelian and Scholastic associations; 'matter' suggests something contrasted with 'mind' or 'life'. The primary naming of *physis* is growth and its first associations are of life and motion. The mere use of it is connected to the Hylozoist doctrine that 'All is alive'. The universe has soul just as the animal body. We must not forget that at this stage of the development of western thinking, *physis* is nearer to life than to matter, self moving because alive.

Anaximander was the first to state a theory of the nature of the world, not only the stuff it is made but also the process of its growth out of the 'limitless thing' into the manifold of definite things. The separation of the elements into their several regions was caused by the 'eternal motion' (a whirling motion of the entire universe, *dyne*) which sifts out the opposites from the primary and limitless 'mixture' in which they will be again merged and dispersed when they perish.

This cosmology thus contains three main factors: a primary stuff (*physis*); an order disposition or structure; the process by which this order arises: *"Things perish into those things out of which they have their birth, according to that which is ordained; for they give reparation to one another and pay the penalty of their injustice according to the disposition of time."* (Diels, 1948).

What is stinking in this Anaximander statement is that the dynamical process is described in moral language: this natural order is considered as ethical. The elements define their provinces following Necessity and Right, as in Homer the Gods are subordinate to a remote power which is both primary and moral: *Moira,* Destiny. The Gods in Homer are younger than the world, they are not eternal, not almighty, and they are limited by the boundaries of *Moira*.

If we turn to the cosmogony in Hesiod, we find that the very division of the world into provinces is made following the supremacy of Moira over the Gods. The cosmogony begins with Chaos, Earth and Eros; then out of Chaos comes darkness and Night, blazing Fire and Daylight. The physical order is guarded by the same powers that punish transgression, the Erinyes or *Moirai* in their darkest aspects. The only difference is that what for Hesiod it had been a matter of faith for Anaximander becomes a theory.

1.2 Totem-clans

Among the Zunis (Mauss,1968) a totemistic tribe of North America, we are told that all the natural objects, and even abstractions, are classified in a solid system based on families and degrees of kinship. In Chinese traditional culture there are similar schemes used as bases for astrology, medicine, martial arts etc.

Archaic beliefs about the nature of the world were sacred (religious or moral) and the structure of the world was a sacred order as well. *In these early phases of social development the structure and behavior of the world were held to be continuous with, the structure and behavior of the human individual and collective mind (society), with reciprocal introjections or projection.*

The first thing to be noted of about the society in the totemic stage is that it is an aggregate not of independent individuals but of groups, which are the basic social units. The totem-clan is defined by a peculiar relation to the totem. The totem is a whole class of objects, generally a species of animals or plants, more rarely a class of inanimate objects; very rarely a class of artificial objects. The totem is something which matters for the society's survival. The members of the totem-clan usually believe themselves to be descendent from a totem ancestor, who is often half human and half plant or animal, a mythical representation which symbolizes the identity of the clan and its totem.

If we take the cosmological scheme of Anaximander, and consider the relation of his primary stuff, *àpeiron*, to the four subordinate elements we observe that it corresponds to the structure of a totemic tribe containing four clans. And we can see that in Anaximander the four elements are organized in pairs of contraries. By this way we are easily brought to the prototype of all social oppositions: the sexual difference. The primal marriage in early cosmogonies is the union of Sky and Earth, later represented in western anthropomorphic religion as the marriage between Zeus and Era. So, the world began from the undifferentiated mass, boundless as the *àpeiron*. This mass separated into two parts, male and female. This description of the origin of the cosmos reflects the social institution of great importance in the totemistic societies: the division of two exogamous phratries: so four is coming from the doubling of male and female groups.

1.2.1 Table of Contraries

The Pythagorean community preserved many traits of ancient societies and it might be not casual that in their system of thinking a special place is given to the Table of Contraries in which as in many exogamous phratries they grouped ten pairs of principles, as many clans, on two columns of opposites. At the head of the two columns stood 'the Even and the Uneven, one of which is unlimited, the other limited, one is female, the other female; while the undivided Monad is bisexed. War and marriage of the opposites is worked out again, in another way, as in other Pre-Socratic cosmologies.

We find, even in Plato, that a sexual character still clings to the great contraries, Form and Matter. 'We must conceive three kinds: first, that which comes into being, second, that in which the first comes to be; third, that from which the first is copied, when it is born into existence. And we may fittingly compare the recipient to the mother, the model to the father, and that which springs into life between them to the offspring.' Even in the desiccated terminology of Aristotle the same representation persists, where he says that one piece of matter can contain only one form, the cause which imposes that form can generate many. 'So it is with female and the male'.

The term 'Nature' (*physis, natura*) has had a long and varied history. Its many meanings can be grouped in the static and the dynamic. Statically conceived, Nature means the total of all existing things, in time and space; the nature of a thing is its constitution, structure, essence. But it has never lost its other dynamic side, the connotation of force, energy. Both these aspects are present in the beginning of interactions between mind and nature: magic.

In a system of totemic human and non human members are not sharply differentiated; neither can be said to control the other. The magical ceremonies are essentially *co-operative*, *sympathetic* and *mimetic*. They try to restore a primary sense of being tuned with the whole, recover the lost *synchronization* between mind and nature. The ritual circumstances exalt the sense of identity of man and nature by identity of behaviors. 'Ecstasy' i.e. getting outside of oneself, implies the existence of a defined self; 'enthusiasm' means being possessed by a power different from oneself. Both these processes belong to mystical religion at a more developed stage. In these primal shamanistic phases there is no separated self, and no need for defined representations.

1.3 Ecstasy *vs.* enthusiasm

We must remember that even the civilized man in the age of philosophy didn't succeed in conceiving anything as immaterial or non- spatial. The *Logos* of Heraclitus, the *Being* of Parmenides, the *Nous* of Anaxagoras, the Love and Hate of Empedocles, are possessed of material properties. Even when the term 'bodiless' or 'incorporeal' (*asòmaton*) appears in Plato it is doubtful how many material properties it negates. A ghost is 'bodiless' but it can be extended in time and space and endowed with some active force, some *mana*, a vehicle of sympathetic interaction.

Just as, to Empedocles, *Philia* (the solidarity of a group) was a mobile fluid running between the particles of the denser forms of matter, the *sympathetic continuum* can only be represented as a fluid which takes the form of the compartments which it fills. These primal groups of the totemic societies are complex and organic and give birth to collective consciousness and religions. The primitive complex of notions (*moira, physis* and *nomos*) were the fixed framework of these collective representations and the personal Gods were acting inside it. So, when the Milesian school left the Gods out of this scheme and tries to deal with natural facts they recovered something from the underneath. Hylozoism simply rose to the level of scientific assertions.

The conception of a primitive living fluid permeated the forms of elaborated constructions. What the Milesian called *physis* has the same origins of what others in other parts of the world call *mana*. These are the roots of modern scientific revolutions, Newtonian mechanics and gravitation, but also represent the persistence of the *subtle body of the subject* inside the first and further steps of science. This subject will regain its uncovered place just many centuries later. Philosophy, after the Olympic Gods, tried to cut its theological superstructures and in this way leaned back on previous animistic cultures. Relations between modern science and animism are neither to be misread nor to be blamed, but they need to be detected to prevent to continuous risk of metaphysics as alienation and reification of abstractions. Metaphysics represents, following (Wittgenstein, 1983) the continuous risk of a peculiar disease in thinking: taking intellect for more real than reality.

1.3.1 Daemons

Daemons are the predecessors of Gods in animism and they survive to the Olympian Gods in Pre-Socratic natural philosophy. They remain impersonal, they consist in will and force without individuality, because they are not the soul of an individual object but of a species or kind (*génos*) to which they are related exactly as the daemon of a human kindred (*génos*) is related to his group. The daemon is an expression of a system of relations. Local fertility spirits differentiated communities of ancient Greece just like in Plato's mythical reign of daemons in the age of Kronos.

This system of poly-demonism, pre-existing Olympian Gods and intermingling post religious thinking in ancient Greece, appears to subsist in Taoism. The *Yang* is subdivided in an indefinite number of souls and spirits called *shen*, the *Yin* into particles or ghosts called *kwei*: they animate objects and man's soul. *Chaos* before it splits into *Yang* and *Yin* occupies the primal place in Chinese pantheon as *Pan-ku*. How does the collective consciousness come to be externalised into a personal daemon or God while on the other hand the group soul gives place to the individual soul?

The king and the hero may have represented transitional forms from the daemon of a group to the individual soul, as the chief was the first individual in the tribal group. The collective authority of the tribe was confounded with his individual will, reinforced with superhuman power. On the other end the chief is a mere depositary of the social authority, a temporary embodiment of power which pre-existed him and will be transmitted to his successors. The same is true for the head of the *gens* or of the patriarchal family. The soul owned by these individuals does not begin or ends with their lifetime, but is *immortal*. The same tradition is present in Hesiod's Age of Bronze: a class of immortals consisting of heroes who fought at Ilion or Thebes. The dream and memory images have been sought to the whole origins of belief in soul and immortality.

The psychology of Homer shows that this *eidolon* may be differentiated from the blood soul, the *mana*. Man in Homer has two souls: his *eidolon* or *psyche* escapes from his mouth at the moment of death. It maintains his shape and sometimes visits his survivors in dreams. The vital force is located in the other soul (*timòs*) whose visible vehicle is blood and is linked to consciousness. In the mortal soul we find the same combination of blood and *mana* which composed the sympathetic continuum of primary magic. As contrasted with the individual *eidolon*, it is less personal and maintains its collective origins. The distinction between these two kinds of soul will be maintained in the following philosophic psychology to the contemporary cognitive science. The *eidolon* soul is the object and subject of knowledge and self-consciousness; the blood soul is the principle of force, motion and intentionality. This distinction will be remarked by Aristotle in the first book of his *Psychology*.

It is also useful remarking the difference between the individual *ker, daemon* and *moira*. The *ker* is an *eidolon* bearing a sinister aspect, and object of fear. It is angry, seeking vengeance, is an *Erynys*. It is also the individual *moira* the limit of his vital force, the negative aspect of his fate. The *daemon* (*genius*) of a person retains a beneficent power and functional *mana*. When Heraclitus, for example, says that a man's character is his daemon says that it is the force that shapes his life from within, and makes his fortunes, not a destiny allotted from the outside. Socrates affirms many times that he is in regular contact with his personal *daemon*.

1.3.2 Olympian and Mystery Gods

This distinction between *eidolon* and *daemon* is someway mirrored by the other distinction between *Olympian* and *Mystery* Gods. The typical mystery god is Dionysus: he represents the life principle of living creatures with its seasonal cycles, birth-death-rebirth. Dionysian cult is based on fecundity, sex, death and re-birth.

The rites were conceived to promote, by sympathetic cooperation with natural powers, the fecundity of Nature, and the synchronization between human and cosmic cycles. This synchronization was promoted by *enthusiasm* (God enters inside the group and the individual) and by *ecstasis* (man rises out of the prison of his individuality and loses himself in the common life of the whole, thus becoming divine). Orgiastic rituals ensure that the passages from human to divine, and from man to nature are always open, and in tune.

Dionysus never became an Olympian and remote god, because his reincarnation and orgiastic cults were always ready to renew these bonds. And his religion is monotheistic, and has many other analogies with other half divine and half human figures like Pythagoras, Buddha and Jesus.

The Olympian god is, like the tribal Hero, owner of a personality, a biography and is detached from the human group. The peculiar rite of Olympianism is the commercial sacrifice in which some benefits must be returned. While each Olympian god in the beginning had a province or a territory of origin and protection; the birth of a pan Hellenic religion meant that gods gradually left their provinces to reach the Olympus, and finally vanish in the sky. Nature deserted by gods is left free for science.

Once earth and heaven are parted Olympian theology is clear that this gap is not to be bridged by Eros. The space between is not to be passed either by mortals and immortals, and the prevailing type of sinner is the one who attempts to pass this gulf, he who seeks marriage with the Queen of Heaven. Olympian theology lays emphasis on justice and necessity of separation of elements. The mixing and "marriage" of elements is an injustice, for example in Anaximander. The mystical schools have a quite opposite view of Eros. Any reader of Plato's Symposium and especially of the discourse of Diotima will know this to be characterised of mystical religion.

One of the speakers in the Symposium, Eryximachus, a pythagorizing physician, sets the two conceptions of Eros in open contrast. In opposition with the Eros of excess, he speaks in favour of the orderly Eros, the principle of binding and *armonìa*. Both of these Loves are present, he says, in the blending of the seasons of the year; and when the hot and the cold, the dry and the wet, are regulated in the relations to one another by orderly Love, and they are mixed by temperate harmony, then they bring a good year and health to men and beasts and plants and there is no injustice. But when the excessive Love prevails there is injustice, destruction by pestilence and other diseases. Such is the physician's point of view: he regards the whole medical art as concerned with establishing the orderly and harmonious Eros between opposites in war on the body.

As soon as the Olympian gods had left their provinces in nature, and the business of the world is realized to precede without them the Ionian natural philosophy can take the place of mystic religion. Materialism can rule, reducing life to mechanical motion, communicated by external shocks from one body to another, discontinuity and discreteness.

1.4 *Physis* and Nature

The Nature of which the first philosophers talk is a metaphysical entity, not merely a natural element, but a power of supernatural life, a substance which is also Mind, Soul and God. This is contained in Thales' words: nature of things is water, the all is alive and has soul, and is full of daemons and Gods. Aristotle reported mostly the first assertion about water, and from him our doxographic tradition is derived; but in cognitive science terms the latter are even more important. What Aristotle reviews in his first book of *Metaphysics* is based on his own distinction of the four primary causes. Modern historians are aware of Aristotle methods, someway parallel to the relation between Socrates and Plato, and rely not only on the material properties of *physis* as a continuous and homogeneous stuff, filling space. They also recognize other physis attributes under the names *Soul* and *God*. Thales declares that the All is alive and has Soul in it (*tò pàn èmpsychon*). This statement accounts for the mobility of *physis*. Its motion and its power of generating things are due to its life (*psykè*) and inner principle of synergy and activity.

Finally this soul substance is declared to be divine: the All is full of daemon and gods, says Thales. This maintains the presence of the superhuman force, *mana*, contained in the notion of magical continuum, and the twin representations of Soul and God. It is a general rule that Greek philosophers describe *physis* as standing in the same relation to the universe as soul does to body: "As our soul is air and holds us together, s a breath or air embraces the whole cosmos" (Anaximenes).

The two vital functions of moving and knowing were less differentiated by early philosophers than by Aristotle. He reports that "the soul is said to feel pain and joy, confidence and fear, and again to be angry, to perceive and to think; and all these states are held to be movements, which might lead one to suppose that soul itself is moved"(Aristotle, 1994). At the same time *Physis* and Soul must be homogeneous. Empedocles said: "with earth we see earth, with water water, with air bright air, ravaging fire by fire, love by love, and strife by gruesome strife".

"Thus those thinkers who admit only one cause and one element, as fire or air, assume the soul also to be one element; while those who admit a plurality of principles assume plurality also in the soul...Those who introduce pairs of opposites among their principles make the soul also to consist of opposites; while those who take one or another of the two opposites, hot or cold, etc. reduce the soul also to one or another of them" (Aristotle, 1994). Aristotle himself proposed the theory of 'like knows like', by maintaining that perception involves a process of assimilation. This philosophic conception reproduces the *sympathetic continuum* of shamans based on the assumption that 'like can only know or act on like'.

This continuum was also the substrate of universal kinship, the primitive form of likeness and interaction. The All is not only alive, but full of gods and daemons. Thales water is pervaded by a divine power capable of moving it. In other words the soul-substance has a superhuman *mana*, a daemonic energy. About the limitless Anaximander's *àpeiron* Aristotle says that it is 'the divine immortal and imperishable'. Of Air Anaximenes says that it is a god and Diogenes of Apollonia asserts that 'what man call air is that which possesses thought, and it directs all and masters all; for just this is, I believe, God and it reaches everywhere and disposes all things and is in everything' (frag. 5). Philosophy maintains pre-existing theological believes on the relation between the God, the human groups and Nature.

These philosophical and proto-scientific theories were not invented but taken over from pre-existing religions. These religions represented in metaphoric ways the primary synchronizing interactions between the inner and outer worlds of mankind.

1.4.1 Art of philosophy

One philosopher embraces a theory or follows another based on his intuition, following his personal attitudes and personality. The charm of these primal philosophies of nature is that each philosopher produces a system of thinking just like a piece of art, dogmatically: "this is the way world has to be". Following an evolved form of sympathetic magic, they give a representation of the object of passionate desire, and the primary representation is mimetic and iconic at the same time. Myth and metaphors mimic emotions and desires in the synchronization with the universal matrix. Differences between control and synchronization are fuzzy, and the notion of causality raise to reach importance to grasp the form under which cause and effect can be differentiated or not. On this basis the emphasis is on likeness, kinship, and material continuity on events not on their temporal succession. The first notion of causality is not temporal, but static, simultaneous and spatial.

The magician knowledge is power over nature, extending the influence of his mana, or the mana of the group. To form a representation of the structure of Nature is to acquire control over it. To classify things is to name them and the name of a thing or a set of things is its soul. To know their names is to have power over their souls. Language is a shadow-soul of the whole structure of reality; it is the most effective and comprehensive tool of human power. Speech is the Logos, which stands to the universe in the same relation as the ritual action to the myth: it explores the chart on the surface of the real. Early science does not distinguish clearly descriptions, explanatory representations and processes of genesis. Geometry has supremacy in abstract mathematics, after its practical origins in land-measurement and it reaches its perfect fulfilment in Atomism.

Plato, in his dialogue Timaeus, introduced the first mathematical model of atomism. He consciously avoided the name 'atom' but he proposed to adopt the Empedocles' four elements as made by Euclidian building blocks, 3-dimensional solids. So, Plato developed a sort of elementary particle physics. The building blocks had the forms that mathematically were considered most beautiful and simplest. Therefore the ultimate root of natural phenomena, in Platonist terms, is not made of natural matter, but of mathematical laws, forms and symmetries (Heisenberg, 1958). *But, how are the forms of living organisms to be explained using regular solids?*

In Antiquity and Middle Ages Plato's approach gained little support because Aristotelian philosophy was considered more apt to understand natural phenomena. Aristotle formulated his concept of balance and equilibrium studying how natural things *function* in reality. Aristotle called the general principle which shapes individuals the *èidos* (form). The basic stuff on which the *èidos* is acting is matter (*ìle*). These principles are not existing in reality, but jus inferred by abstraction. Matter is also defined by its potential to be formed (*dùnamis*) and the result in reality (*energèia*).

1.4.2 Language of philosophy

Aristotle's approach to language, which is obviously inherent to mind nature, is someway in contradiction. In *De Interpretatio he asserts that "what is happening in speech is a symbol of the affections in soul, and he had the idea that there could be a strict parallelism between phrase structure and represented processes. In this way he was following Heraclitus who clearly believes that the name is a natural property of the thing: the same type of archaic attitude towards language showed by tribal cultures and infants. The Eleatic school opened the way (here again) to a scientific study of language by recognizing that the Universal Being is independent from linguistic habits, which are "imposed by conventions" and "not true": the 'dog' word doesn't bite! Relations between words, thinking and things are more complex and complicated.*

Plato exposed in the *Cràtilus* dialogue an evolving theory: he proposes that *each word is a special tool* (*òrganon*) forged by mankind to evoke a thing. But the Platonist solution is that a thorough study of language is not so important to understand the Real. Aristotle maintains something of his master's position: linguistic forms are generated by men, following an institutional convention in human societies. He also asserts that the study of language is fundamental in building science, and a well ordered town. If Aristotle's conception would be true, the study of language would be the golden door to know both reality and mind. This was the idea of Wittgenstein in his first *Tractatus*.

Augustine is catching the strong connection between linguistic expression and the body, but he risks (and really adopts) a referential theory of language: every word has a meaning, and the meaning is a real thing. "When they (my elders) named some object, and accordingly moved towards something, I saw this and I grasped that the thing was called by the sound they uttered when they meant to point it out. Their intention was shown by their bodily movements, as it were the natural language of all peoples: the expression of the face, the play of the eyes, the movement of other parts of the body, and the tone of the voice which expresses our state of mind in seeking, having, rejecting or avoiding something. Thus, as I heard words repeatedly used in their proper places in various sentences, I gradually learnt to understand what objects they signified; and after I had trained my mouth to form these signs, I used them to express my own desires." (Augustine, Confession 1.8). Wittgenstein, criticized his own young positions, presented in his *Tractatus* (Wittgenstein, 1961) starting from Augustine, in his later *Philosophical Investigations* (Wittgenstein, Ludwig, 1967).

Augustine had also an interesting theory of time in mind processes: "What is time? If nobody asks me I know it; if I try to explain I don't know it". He realizes that neither past nor future, but just the present really *is*. This exploration in various dimensions of (his) subjectivity made him anticipating the Kantian theory of time and Descartes' *cogito*. In his *Soliloqui* he writes: "You want to know, but do you know if you are? I don't know. Where you come from? I don't know. Do you feel single or multiple? I don't know. Do you feel moved? I don't know. Do you know you are thinking? I know."

This position was opposed in late Middle Age by the Nominalism of Ockam, inventor of the famous "*epistemological razor*". He stressed that names, like *homo*, *animal* etc. are not to be considered really existing out of our soul: they are just linguistic forms used by our mind to establish logical relations.

Saint Thomas, Aquinas, had a real and deep knowledge of Aristotle because his friend Wilhelm from Moerbecke provided him fresh translations from Greek, and he wrote also some commentaries. Some philosophers assert that Averroè is more loyal to Aristotle than the Aquinas, who is a more faithful servant of the Roman Church, but he developed some original thinking about soul. In the *Summa Contra Gentiles*, book two, he recognizes that soul is united to the body, and it is the form of the body (just like in Aristotle).

All the soul is present in each part of the body, and animals have souls, too: though their souls are perishable, and not immortal as man's soul. Intellect is a part of each man's soul, and there isn't a universal intellect participated by all men, as Averroè was suggesting. Soul isn't transmitted by sperm, but is created from nothing for each man. If the newborn comes outside of a regular marriage, it seems that God is participating to a sin, and we also could ask about the transmission of the Original Sin…but we won't find answers for this in the Aquinas' works. It is more important that the Aquinas re-introduced Aristotle in Western thinking and particularly his theory of *active perception*. This means that *organisms learn about the world by actions in the world. This activity incorporates at the same time perception, action, learning and change (in the subject and the environment)*.

It is called *intention*, from Latin word *intendere*, which means 'to stretch forth', and equally a change both in the self and the object, by defining a goal, experiencing an action and learning from the consequences. Aquinas, though criticizing Plato, just like Aristotle, differentiated himself also from the latter by proposing that interactions between mind and world aren't perfectly symmetrical. Intentions are realized in actions changing the world and the sensory consequences enable the body to change itself in accordance with the nature of the world.

There is assimilation, *adaequatio*, in which the body changes its own form in order to become similar to stimuli relevant to the emerging intents within the mind. There is an interesting variation on Plato's cave proposed by the Aquinas. The cave becomes a tent in which an observer infers what is happening outside from the patterns of reflected light and the movements of the tent walls. We remember that in Plato's cave light and forms coming from the outside are caught by the senses from shadows on the passive walls, while in the Aquinas tent forms are created from the inside by reaching a similitude with the outside.

For example, when you shape your hand to hold a glass of water you don't transfer geometric shapes from the heaven of forms to your brain, but simply shape your hand to the form of the object: no two glasses are identical, and exactly in the same context of action. In this way Aquinas thinking becomes so modern because "the body and the brain are open systems with through-put of matter, energy and information, but the uni-directionality of perception makes the fabric of meaning a closed system" (Freeman, 1999).

The Aquinas anyway, in being a soldier of the Roman Church, was also re-evaluating Aristotle's theories of language, in contradiction with his own theories on perception and intentionality. Against sceptics, either naturalists or sophists, language is the writing of soul. It allows a reliable transcription of psychical data and their reliable transmission to the outside. Words and data (inner or outer objects) have stable and objective relations. Language, in Aquinas approach, becomes the perfect weapon against sceptics, and in this way it has to be a perfect language, but it loses a dynamic connection with intentionality.

1.4.3 A multiplicity of languages

Dante Alighieri was de-structuring this view in proposing a *multiplicity of languages*, in favour of vulgar against Latin. He was convinced that Latin can be a perfect language, the grammar *par excellence*, the langue of science; but he doesn't believe Latin is good enough to express emotions, the everyday-life world. Vulgar is the right language, as it is spoken by everyday people, and written in prose and love poetry. In this direction it is always useful to stress that he wrote the Divine Comedy in Vulgar, and not in Latin. This opened new doors for perception, jut like the new geographical discoveries were going to do a few years later.

Leonardo da Vinci, in his Notebooks was thoroughly studying the inner and outer turbulences of deluges, while at the same time, tries to represent the Neo-Platonic perfections on in the symmetries of the characters in his paintings. All Leonardo's work is marked by his fragile and extraordinary balance between order and disorder. Anatomical studies and sketches on floods, perfect portraits with fractal landscapes.

Also new geographical discoveries contributed to open doors for new perceptions of the world, and a new crisis for the holistic tradition of Aristotle, Avicenna and the Aquinas.

Galileo will affirm, against the Roman Church, that phenomena discovered in the new, deterministic, physics have no names in historical languages and are in search for new names in mathematical language. The dialectics between holism and determinism are parallel to the dialectics between natural and artificial languages.

Linnaeus is a living example of this process and its inner contradictions because at the same time invents a new language for his classifications, and proposes that these names are coming from God, creator of all the animal and vegetal species.

1.5 Sum, ergo cogito

Doubt seems ready to become the golden path to knowledge and Descartes is ready to follow it. In his *Meditations* (Descartes, 1998) he gets to his primal intuition of the human mind by methodically doubting everything. Methodical doubt is the paradoxical and perfect tool to find, in a process similar to the sculpting *Ars in Levare*, the perfect truth in what remains as indubitable. "But what then am I? A thing which thinks? It is a thing which doubts, understands, affirms, denies, wills, refuses, which also imagines and feels" (cit). But the difficulty arises to explain interactions between mind and body, just as they are defined as separate entities: how can an un-extended mind (*res cogitans*) interact with the extended body (*res extensa*)?

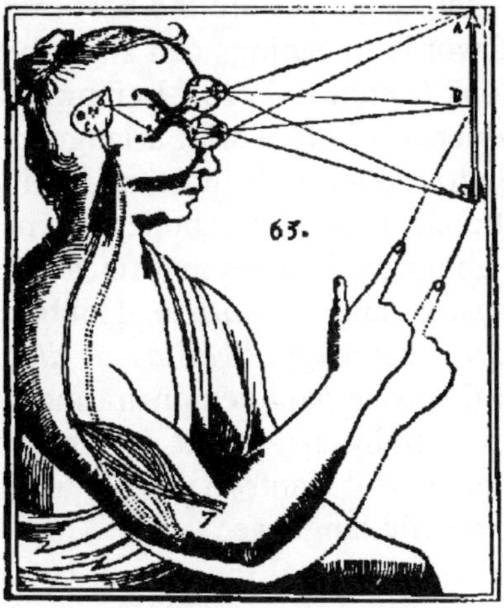

From Descartes' *Dioptrique*

His basic insight was rich of implications he didn't explore. Being and existence are different, for example, just as thinking and being. He also wrote in his book *Passions of the soul* about all the emotional states like fear, love etc. attempting the same enterprise as Darwin many years later. Descartes in his study of passions proposed a mechanical model based on the actions of "animal spirits" on the body.

Descartes made several errors, mostly having considered thinking as an antecedent of being. Probably it is more correct to say "*sum, ergo cogito*": as thinking would require a container. Anyway, he placed all the basic questions and prepared the land for the development of modern cognitive science.

Leibniz also offered a contribution in this direction which is not well understood. Besides his discovering of differential calculus, done at the same time with Newton, he was conscious of the impasses embedded in the mechanistic approach to mind. That's why he proposed the *monads* which can be seen, nowadays, as forerunners of the CAS (Complex Adaptive Systems) and artificial life approaches to complex systems. In his third book of the *New Essays on Human Intellect* he recognized that each language finds its peculiar organization in syntax and morphology: speech is a way to mould experience. At the same time (in tune with Locke and Vico) a language reflects history and tradition but also conditions habits and culture.

"I truly believe that languages are the best mirror of human spirit and a rigorous analysis of words' meanings can show better than any other study how intellect functions (Leibniz, 1996). Leibniz considers talking as a way to organize the experience of the world: a language reflects the history of a nation and is capable to mould its beliefs and habits. He defined a project to construct an international artificial symbolic system which should have taken the place of natural languages in the scientific world.

In his *De Corpore* and *Leviathan* Hobbes proposed a theory of knowledge as based on language: "*veritas in dictu, non in re*". And John Locke proceeded on the same path demonstrating how possess of words and learning can be basic conditions in the development of general ideas and theories. Real knowledge, if it wants to reach a universal value, must become independent from a specific language.

Hume in criticizing traditional theories of causality (*Treatise of Human Nature*) developed a theory of the mind-body set. If, according to Descartes rationalism, human mind rules the body like a monarch, for Hume there is no separate substance but just a self organizing field of emerging and disappearing patterns excited by associations and sense impressions. This organization can be compared to the flexible structure of a democratic society, animated by self-organized groups and parties.

Kant tried to synthesize rationalism and empiricism building a rich philosophical construction. Recognition doesn't come from a *tabula rasa* in the brain, but from an active process which produces models of the world by *a priori* categories. These cognitive schemes organize the mass of experienced data, but the organization has a limit on the outside existence of the unknowable *Ding an Sich* (the thing in itself). The dynamic process of knowledge becomes an activity regulated by two limits: categories and things-in-themselves. The partial independence, or relative dependence, between mind and reality can define their negotiations in this mutual agreement of relative freedom. In his *Critique of Judgement* (Kant, 1987) Kant, in discussing the nature o f living organisms, asserts that they are complex, self-reproducing and self-organizing.

A living organism is different from a machine in the sense that each part exists for, and produces the others. "We must think every single part as an organ which produces the others (and reciprocally is produced by them)...Just then and for this reason, this product, as an organized and self-organizing being, can be defined as a natural goal". Kant was the first to use the term *self-organization* to define living organisms, and he used this term in a way so similar to the current field of dynamic systems theory. On the relations between language and perception he writes in his *Principia Analitica*: "At the basis of our pure sensitive concepts there aren't pictures of objects, but schemes (patterns). No triangle image could be adequate to the general concept of triangle. [...] The triangle *pattern* cannot exist elsewhere than in thinking, and it means *a rule in the synthesis of imagination* [...] This *schematism* in our intellect, related with phenomena and their immediate form, is a technique (*Kunst*) hidden in the depth of human soul" (p. 100). This passage is very interesting, but elusive about the role of natural language, and the opinions that Kant had about the precurring works of Locke and Berkeley.

Hamann wrote his *Metakritik*, a contemporary and constant criticizer of Kant, to answer that human reason in delimiting itself can never be *pure* but relies in natural language habits which can never be abandoned. *Meanings aren't just symbols, but also schemes and condensation nuclei for our knowledge.*

1.5.1 Principium individuationis

Friedrich Nietzsche (2000), in his first important book, The Origins of the Tragedy offers a new synthesis on the dynamics of mergence-emergence. He proposes that art, here represented by the classical Greek tragedy, take origin from the meeting of *Apollo and Dionysus, as morphogenetic principles*. It isn't casual that the two gods often shared the same sanctuaries, as in Delphi and Delos. They represented complimentary principles, and we can also consider Apollo as the ambassador of Olympic gods to Dionysus. While Apollo represents visual arts and perception, Dionysus represents non visual arts (music) and the whole non visual perception (hearing, tasting, smell) in his orgiastic cults: "until they finally, through a marvellous metaphysical act seem to pair up with each other and, as this pair, produce Attic tragedy, just as much a Dionysian as an Apollonian work of art."

Nietzsche reminds that Schopenhauer (1907) said about the man trapped in the illusionary veil of Maja: "As on the stormy sea which extends without limit on all sides, howling mountainous waves rise up and sink and a sailor sits in a row boat, trusting the weak craft, so, in the midst of a world of torments, the solitary man sits peacefully, supported by and trusting in the *principium individuationis* [the principle of individuality]". Apollo is the incarnation of the *principium individuationis*, while Dionysus represents a *principium fusionis* [principle of fusion]. Art, in Nietzsche vision, represents "This reconciliation is the most important moment in the history of Greek culture. Wherever we look the revolutionary effects of this experience manifest themselves. It was the reconciliation of two opponents, who from now on observed their differences with a sharp demarcation of the border line between them and with occasional gifts send to honour each other."

In the Dionysian dithyramb, as in the shaman's songs, and tribal dances, "man is aroused to the highest intensity of all his symbolic capabilities. Something never felt before forces itself into expression—the destruction of the veil of Maja, the sense of oneness as the presiding genius of form, of nature itself. Now the essence of nature must express itself symbolically; a new world of symbols is necessary, the entire symbolism of the body, not just the symbolism of mouth, face, and words, but the full gestures of the dance—all the limbs moving to the rhythm. And then the other symbolic powers grow, those of music, rhythm, dynamics, and harmony: all with sudden spontaneity.

1.6 Emergence *vs.* Fusion

Emergence is a universal phenomenon that can be defined mathematically in a very general way. This is useful for the study of scientifically legitimate explanations of complex systems, here defined as hyper-structures. A requirement is that the observation mechanisms are considered within the general framework. As we defined in the introductory remarks in this chapter, emergence is related to the ways we consider the structure of a given system.

Living as cognitive beings in a world of stability and change, we permanently face known and unknown situations, old and new phenomena - conversations, happenings, shifting perceptual patterns of the world. We usually cope with the complexity of ordinary life in a way so easy and by such seemingly simple patterns of behaviour - when seen from the higher level description of ordinary language - that it is astonishing to realize the underlying complexity of this `problem solving behaviour' from a scientific point of view, i.e., when we try to relate descriptions of the body as a complex system and its environment (as given by physics, biology and cognitive science) to the life and actions (as experienced) of our daily world.

In the so-called sciences of complexity (e.g., non-linear dynamics, theoretical biology, complex adaptive systems, artificial life, artificial intelligence, cognitive science), "complex" phenomena, such as the appearance of life on Earth, the evolution of new species, or the structure of cognitive thought, are often considered as instances of some emergent higher-order structure that may be explained by the lower-level dynamics generating the collective behaviour or emergent property of the system in question. Can the emergence of real new properties in complex systems really be explained? Yes, if the sciences of complexity offer new important insights, theories, and methodologies for dealing with complex, higher-order phenomena. Perhaps the very idea of scientific explanation as a strictly deductive argument should be reinterpreted, and explanations seen in a more dynamic and context-dependent setting. *Explanations are emergent structures*.

1.6.1 Hyper-structures

We have to pay attention to a new general framework for description of higher-order structures, *hyper-structures*, which includes the mechanisms of observation, and which eventually allows for self-generation in such systems of new observational frames. New observers may also emerge in the system (Baas, 1994).

When it comes to such questions as the very generation of explanations, the discovery of new structures or principles, the appearance of new insights to the inquiring mind - i.e., the emergence of structures that can be said to explain a phenomenon - very little has been said. The feeling of having explained a phenomenon cannot simply reduce to the fulfilment of some algorithmic procedure by which - given a general law and some initial or boundary conditions - you can arrive at the phenomenon to be explained by executing this procedure. Of course, one can argue that it is the very laws of nature that has the explanatory power in physical science, so any research programme should attempt to discover such laws. However, in understanding complex systems, the challenge is often to understand the consequences of these laws on other levels of organization, i.e., to understand the history and complexity of the boundary conditions. *Understanding a system means creating some kind of generalized resonance, by this way there is a link between understanding and change: for both parties, system and subject.*

Traditional deductive notions of scientific explanation have frequently been criticized for being too crude and reductive to account for complex mental phenomena. Hence, in a somewhat romantic vein, the mathematician (Spencer-Brown, 1994) declared that:

"To explain, literally to lay out in a plane where particulars can be readily seen. Thus to place or plan in flat land, sacrificing other dimensions for the sake of appearance. Thus to expound or put out at the cost of ignoring the reality or richness of what is so put out. Thus to take a view away from its prime reality or royalty, or to gain knowledge and lose the kingdom."

The pure dilemma of reductionism and holism is an illusion: we can gain knowledge and retain the richness of a world of emergent structures.

1.6.2 Emergence in hyper-structures

In the recent years the notion of emergence has been studied extensively, but often without making precise what emergence means. We will here use emergence in the general sense defined by Baas (1994). The crucial point in this definition is the notion of *an observer* - in a very general sense. In the study of complex systems one often sees that a collection of interacting systems shows collective (nonlinear) behaviour. This is intuitively what we understand by emergence. In order to study the phenomenon further it may be useful to introduce a more formal framework as follows:

$$\{S\}\, i \in I$$

let it be a family of general systems or "agents". Let Obs^1 be "observation" mechanisms and Int^1 be interactions between agents. The observation mechanisms measure the properties of the agents to be used in the interactions. The interactions then generates a new kind of structure

$$S^2 = R\,(S^1{}_i\,,\, Obs^1\,,\, Int^1)$$

which is the result of the interactions. This could be a stable pattern or a dynamically interacting system. We call S^2 an *emergent structure* which may be subject to new observational mechanisms Obs^2. This leads to:

$$\text{P is an emergent property} \\ \Updownarrow \\ \text{P} \in Obs^2\,(S^2) \text{ and P} \notin Obs^2\,(S^1{}_i)$$

The observational mechanism may be internal or external.

Examples:

- Coupling of dynamical systems. (control and/or synchronization);
- Large collections of objects get new behaviour (like in phase transitions);
- Functionality of bio-molecules in cells.
- Consciousness is not a property of individual neurones.

With the interactive help from the coupling each one may perform tasks which none of them could do separately. Hence we get a *second order agent* (CS) which again may serve as a master in a new context. This is related to *control and synchronization theory.* Consciousness is not a property of individual neurones; it is a natural emergent property of the interactions of the neurons in nervous system of the body in an environment. It makes a structure that is related to lower level interactions as well as higher level thoughts, and it represents a new observational mechanism of the entire system.

General causes of emergence would be: non-linear interactions, large collections of limits, open-ended evolution in heterogeneous environments, and context-dependence of properties in complex systems...

Depending on the case, the $Obs^2\,(S^1{}_i)$ ^{may} have elements, or it may be an empty set if for instance *Obs* does not or cannot observe anything.

1.7 Interactions as morphisms

We think that the present framework is quite useful for analysing the nature of emergence - in particular the dependence on observational mechanisms. In order to put this into a more mathematical framework category theory is very useful. *We let the systems be represented as objects in a category and the interactions as morphisms*. A complex system with interactions is then represented by a diagram. Since the morphisms are represented by arrows, this may be viewed as a process oriented scheme.

The observational mechanism may here be thought of as functors. In categories we may form direct limits ($\varinjlim$) of diagrams, and in this case the $\varinjlim$ may be thought as both an observation functor (*Obs*) and a complexification process (*R*).

It is always impossible to deduce the properties of the whole from its constituents and the observational mechanisms. Thus, within the general framework proposed here, one must distinguish between two different kinds of emergence:

A. *Deducible emergence.* There exists a deductive or computational process or theory *D* such that $P \in Obs^2(S^2)$ can be determined by D from (S^1_i, Obs^1, Int^1). This is structural or ontic emergence.

B. *Observational emergence.* If P is an emergent property, but cannot be deduced as in (A). This is a phenomenic emergence.

Examples of deducible emergence include some compositional structures in nonlinear dynamical systems, phase transitions, non-triviality of complexity of manifolds in topology. In these cases, the various properties can be decided by well-defined procedures, so *Obs* can be seen as instantiating an algorithm leading to a set of properties. Examples of *observational emergence* include `Gödel sentences' in a formal system (the *Obs* is the truth function; cf. Gödel's theorem), and the property of membership of the Mandelbrot set and most Julia sets.

Furthermore, it was indicated that the eventual semantic non-compositionality of a language would imply that the meaning of sentences in such a language was observationally emergent. In formal string systems, compositionality can be defined as the condition that there exists a computable function *F* such that

$$M(s) = F(M(s1), ..., M(sk)),$$

where the *s*'s are the syntactical relations and *M* is a semantic "meaning function". If we take *Obs* as our *M*, then *compositionality* amounts to deducible emergent meaning, whereas *non-compositionality* would imply the existence of observationally emergent meaning. As opposed to formal systems, the meaning of sentences in natural language are often context-dependent, i.e., the meaning of a sentence (such as "He saw the light") is not simply deducible from the meaning of its components, but dependent of the larger discourse in which it is placed. Thus, *observationally emergent meaning* constitutes a problem for the semantic analysis in the Tarski tradition (which is set theoretical, truth functional and compositional), as well as for classical AI that follows `the formalist motto' of deducing semantics from syntax.

In spite of the existence in mathematics of a large set of yet undecided statements or conjectures about mathematical properties that may exemplify either deductible or observational emergence, it is a general observation that the most transparent and clear-cut cases of the two kinds of emergence are found within the logical and mathematical domain. Nevertheless, the distinction may also have considerable potential value in the empirical sciences, for instance in the ongoing discussion of reductionism and the explanatory relations between various theories. To give an example, it is a sensible intuition that the *autonomy* of biology in relation to physical science is, so to speak, grounded in the observational emergence of specific properties of biosystems, such as the self-reproduction of living cells. For theoretical biology it is of interest to determine in what sense such observational emergent properties, as studied by Artificial Life models, can or cannot be considered as real instantiations of living entities. The same can be said for mind sciences.

1.7.1 Autopoiesis

One may go even further by using the concept of a living system as an *autopoietic* system (Maturana & Varela, 1980), i.e., a system organized as a bounded network of processes of production, transformation and destruction of components that regenerate and realize the very network and its boundary and thereby constitute it as a unity in this new (autopoietic) space of relations. This property can, we suggest, be understood as the observational emergence in the physical space of systems that realize their own self-production, boundary and self-observation (through the boundary's distinctive or selective property, which is based on molecular recognition reactions by the membrane-bound proteins), and thus being `cognitive' in this primitive sense. If this is true, an autopoietic system is autonomous because it realizes observational emergence of itself as an observer. *Hyper-structures* are multi-level emergent structures.

A hyper-structure of order N is given by

$$S^N = R(S_{i_{N-1}}^{N-1}, Obs^{N-1}, Int^{N-1}, S_{i_{N-2}}^{N-2}, \ldots)$$

extending the construction in the definition of emergence. This is a cumulative structure, not necessarily purely recursive.

The client/(ob)server situation may easily be extended to a hyper-structure of the *N*-th order clients or (ob)servers. Within the living realm, one can observe hyper-structures as realized by multi-cellular organisms and the community structure of ecosystems. Consciousness may be seen as a hyper-structure of mental representations embodied in the central nervous system and capable of self-observation and self-interaction.

Hyper-structures may be thought of as an organizational scheme or design principle. The role of *Obs*- which may also be an interacting environment - may also allow for evolution to be incorporated.

The point is to combine the notion of *emergence and hierarchy* into the notion of *hyper-structure* which implies also *fusion*.

Wherever a problem, a situation, or a structure can be organized into a hyper-structure, this has an explanatory power. Cognitive in the sense of Maturana and Varela where a cognitive domain is the entire domain of all interactions in which an autopoietic system (an organism) can enter without loss if identity. We think that the framework proposed in this paper allow for an interpretation of the theory of autopoiesis that adequately integrates evolution and emergence of autonomy into the theory.

1.7.2 The Ob-server

As already emphasized, the notions of *emergence and hyper-structure depend critically on an ob-server*. We think that this is natural in as far as phenomena has to be explained through observational facts (even in the abstract sense of facts).

An example: When we model phenomena in thermodynamics by statistical mechanics we may view temperature and pressure as emergent properties and phase transitions as emergent phenomena. These are parts of collective behaviour which in many cases are well understood in a reductionistic sense in physics. However, temperature has been introduced as a formal parameter ad hoc. A deeper level of explanation - not yet reached - would be to see the *laws* of thermodynamics derived from the laws of Hamiltonian mechanics, i.e., such that temperature would *emerge*.

More informal examples may be the emergent properties of hyper-structures of biological and social systems which are deeply dependent of the existence of observers intrinsic to the system. As argued above, the single cell with its membrane bound proteins constitute an observer of some aspects of its immediate environment (other molecules that can be recognized as signals or nutrition), and multi-cellular organisms depend critically on inter-cell signalling. On a higher level, most if not all institutions of a human society have self-observing mechanisms (e.g., evaluation and assessment in research and production), as well as mechanisms for observing other institutions.

Take for example the topological ways complicated surfaces like manifolds are being glued together from elementary pieces. New topological and geometrical properties occur, but may often require *sophisticated observational functors like co-homology theories* to be detected and described. Furthermore, in a knot where is the knottedness? It is a global property, having no meaning locally. Or in a Moebius band: where is the *twist*?

Category theory is the mathematical language suited for discussing structures in general. A category consists of

- Objects: X, Y, ...(like spaces, groups, algebras, ...)
- Morphisms: X Y represented by arrows (functions, interactions)

To each pair of objects we have a set of morphisms (X, Y). A category is closed under morphism composition

$$x \xrightarrow{f} y \xrightarrow{g} z$$

$$g \circ f$$

meaning that if f and g are morphisms, so is also $g \circ f$.

We can form networks of morphisms as commutative diagrams and construct their categorical limits.

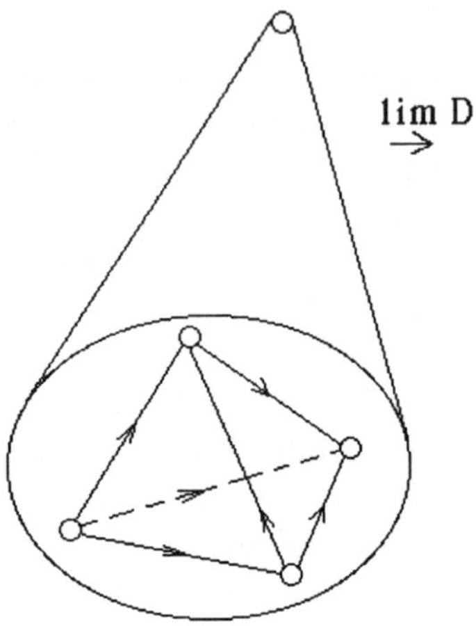

$\varinjlim$ D is an object which is a kind of synthesis of D and which is called the direct (or co) limit of D. Limits often play the role of emergent structures.

The *limit construction* may be viewed as a *complexifying process leading to new emergent properties* in many cases according to some observational mechanisms. Specifically, if

$$D = \xrightarrow{\lim} D_i \text{ in } \mathcal{C}$$

F a functor:

$$F: \mathcal{C} \quad \mathcal{E}$$

$$F: \quad \mathcal{C} \rightarrow \mathcal{E}$$

say F(Di) = 0 trivial object in $\mathcal{E}$, but if F(D) is a non-trivial object in $\mathcal{E}$, we may view F(D) as an emergent structure or property of the diagram.

1.8 Reductionism and non-reductionism

Within the framework of category theory we can beautifully illustrate the differences between reductionism and non-reductionism. We appeal here to the figure:

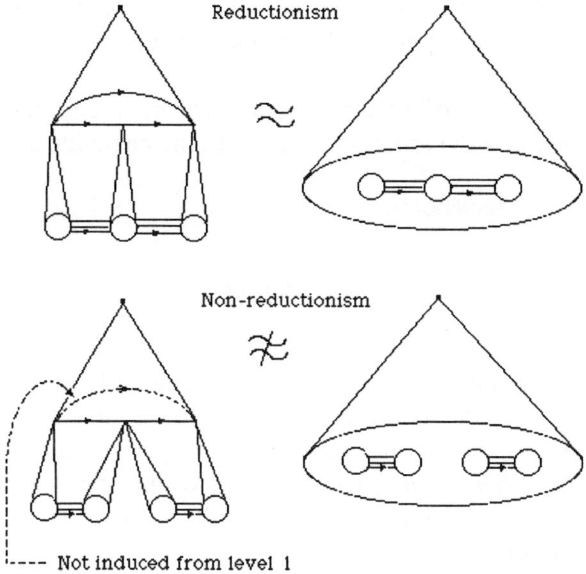

The circles represent diagrams in a category, the horizontal arrows are morphisms, and the cones are limits.

Normally we mean by an explanation of something that there is a deductive procedure, an algorithm, leading us from something basic, well-known to the new thing. In essence this is reductionistic explanation as formally illustrated in the figure. An observer just singles out in a situation what is to be explained. Therefore *deductive emergence is basically a kind of traditional reductionistic explanation, but with an observer as an additional enrichment*. Non-deducible or observational emergence is more subtle. In this case there exists no algorithmic explanation. As we see in the figure in the case of limits a two level structure is needed to "explain" a structure. It cannot be explained or understood only from the primitive level. At level 2 new phenomena emerge (in this case caused by the composition properties in categories), which are needed at the next level. Even if there is no algorithm from the bottom up, still we would claim that this construction or figure is a perfectly good explanation - a second order emergent explanation ($\varinjlim$ may be thought of both as a construction and as an observation). The iterated limit construction is just a special case of a hyper-structure, which is a multilevel emergent structure. Therefore we will say that we have higher order – *N*-th order explanation of a phenomenon- wherever it occurs as a result of a suitable *N*-th order structure.

In our discussion of hyper-structure we have suppressed time, which should be added in a dynamical picture. But the basic ideas remain the same.

We may think of the scheme

$$\frac{S_1,\ldots S_n}{S} \ \mathrm{rel}\ (\mathit{Int},\ \mathit{Obs})$$

meaning that *S* is an emergent structure, as an inference rule. A property observed in *S*, but not in the *S*'s could then be said to have an emergent explanation. This extends to more levels and hyper-structures in general, and could lead to a *dynamical process logic* which we would call *emergent deduction*, which would be useful in order to explain and reason about emergent phenomena.

The important - and new - point is that emergent deduction is not only a *syntactic* process, but combines *syntax* and *semantics* in the reasoning and explanation. After all this is how much human reasoning occurs - for example mathematical thinking.

In science we often search for fundamental laws that cannot be reduced as logical consequences of something else. But just as important as fundamental laws are it is vital to understand their consequences or what may be called *emergent laws*. These would be laws (or in general: organizational and dynamical principles) that emerge from the regularities in collective behaviour of structures - the laws that emergent structures obey. A striking example would be superconductivity. At the biological level, it could be natural selection of genetically based self-reproducing entities.

In forming hyper-structures there will be several levels of emergent laws, and they will provide explanatory tools in the sense of emergent explanation.

1.8.1 Explanation and insight

On the subjective level, it is a common experience of people to have sudden "flashes of insight". They involve the emergence in consciousness of a possible solution to a problem that for long have occupied the working mind (its conscious as well as subconscious parts). The sudden "flash" is experienced as the appearance of the solution as a new structure, that in a sense can be observed at once, but nevertheless has to be worked out deductively in great detail and tested formally before it can be trusted. This experience has been vividly described by Poincaré (1929).

The appearance of a new idea or a new orderly structure may often be similar to the detection of a pattern - or a special "short description", rule or algorithm, which may generate the more complex situation that was the starting point. Of course, such experiences are not confined to mathematicians; a very common feature of consciousness in general seems to be the property of "overflow" (as described by Searle) in which whole and complex situations experienced in personal life suddenly are seen in a new light, from a new point of view as it were, as if a higher order observational mechanism has accessed properties of new structures as generated through the interactions of previous experiences. We propose that the characteristic phenomenon that conscious states refer *beyond* their immediate content (as seen in the overflow phenomenon, in the so-called "Aha! Experience", and in the well-known Gestalt shifts between figure and ground) is deeply related to the hyper-structured constitution of consciousness.

1.8.2 The emerging mind

To explain the mind without losing the very phenomenon, the explanation must not, to quote Spencer Brown again, "take a view away from its prime reality". The traditional conflict between objective explanations of consciousness (the attempt to discover its law-like relations with other things) and subjective understanding of consciousness (i.e., of its intrinsic qualities, phenomenal aspects, "what it is like", etc.) may be transcended if we integrate the following ideas into the general framework of hyper-structures:

If laws of nature are a kind of algorithms giving us highly compressed descriptions of the phenomena we can observe in nature, the existence of observationally emergent properties (that cannot be computed from the lower level algorithms) shows the necessity of reformulating the notion of lawful explanation so as to encompass the historical generation of hyper-structures.

Any scientific explanation has objective as well as subjective aspects. As objective, the explanation can be publicly stated, explicitly and precisely (eventually in a formal mode), it can be critically discussed (and eventually tested), and it can generate adequate representations of the phenomenon to be explained ("explanatory power"). As subjective, the explanation must be understood and make sense to a competent observer; it must be fully or partly accepted as a genuine explanation (i.e., as one that generates the phenomenon); and it must thus mentally reflect the emergent structure of the phenomenon to be explained.

Not all of reality is objective; some of it is subjective and conscious (though still a completely natural phenomenon). For the objective reality, one can often maintain the traditional view of observation as something in which the observer is distinct from what is observed. For some aspects of the subjective reality, this model of observation cannot always be maintained.

Hyper-structures can be found existing in objective as well as in subjective reality.

The methodology of a science of consciousness must concern itself not only with objectively observable behaviour (from the third-person perspective), but also with what has been termed first-person point of view, or what we will call the observer observing the emergence of new structures from within the system, as exemplified in mathematical discovery or in other kinds of sudden flashes of insight.

In certain living systems with mind, at a high level new phenomena such as consciousness emerge as a construction process, in which *Int* and *Obs* occur at several levels, and in which (at least some) observation mechanisms are intrinsic to the system. Even though there is probably no algorithm from the bottom up, such a self-construction is a perfectly good explanation - a high order emergent explanation, open for further analysis.

The approach presented here does not view the mind as self-transparent, nor does it invoke self-introspection as a privileged observation mechanism; rather, it relaxes the exaggerated expectations and quixotic hopes for some completely different sorts of explanations for conscious phenomena. Even if some aspects of consciousness, such as *qualia*, may constitute a particularly hard problem, other aspects may be explained within a refined framework of emergent hyper-structures. Without ignoring the richness of the phenomena of life and consciousness, we can still continue to work out better explanations.

After the current studies on neural plasticity we can also say that brain emerges from mind, because it is moulded by mental events. As in the Moebius band, container and content in the mind-brain relationship can be reversed in their reciprocal function depending from the perspective of the observer. That is, an increase of knowledge about the way certain effects are obtained may reveal that certain effects are decomposable into the effects contributed by subcomponents of that system.

We have to consider both the whole (decomposable), its parts (fusible), an observer function. There is also a possibility that the observer is observed by the whole, in this way becoming a whole object, and/or mirrored.

Chapter 2: Semiosis

...information is a difference
that makes a difference.
Gregory Bateson

2.1 The empire of signs

If we can say that the 20th has been the century of energy, we can also say that the 21st is going to be the century of language. Language, as the core of most advanced research, is pervading every form of scientific and technological knowledge on the edge of innovation: from information technologies, to biology and mind sciences. As we defined in the *Preface*, language is the red thread in this exploration. *Language, in its subset of Information Theory, combinatorics, is also the interface between biology and physics, and between mind and brain.*

Information Theory is embedded in the core of Semiotics, the broader discipline concerning any kind of sign and communication. Its beginning, though implicit in the Pre-Socratics, starts officially with John Locke (, 1694) who coined the term (semiotics), as a Greek neologism. His remarks at the end of Book IV of the Essay virtually amount to a suggestion of the program that Peirce proposed for himself but never completed:

"The third branch [division of the sciences] may be called *semiotikè*, or the doctrine of signs; the most usual whereof being words, it is aptly enough termed also *logiké*, logic: the business whereof is to consider the nature of signs the mind makes use of for the understanding of things, or conveying its knowledge to others. For, since the things the mind contemplates are none of them, besides itself, present to the understanding, it is necessary that something else, as a sign or representation of the thing it considers, should be present to it: and these are ideas. ... To communicate our thoughts to one another, as well as to record them for our own use, signs of our ideas are also necessary. Those which men have found most convenient, and therefore generally make use of, are articulate sounds. The consideration, then, of ideas and words as the great instruments of knowledge makes no despicable part of their contemplation who would take a view of human knowledge in the whole extent of it. And perhaps if they were distinctly weighed and duly considered, they would afford us another sort of logic and critique than what we have been hitherto acquainted with… All which three, viz. things as they are in themselves knowable; actions as they depend on us, in order to happiness; and the right use of signs in order to knowledge, being *toto coelo* different: they seemed to me to be the three great provinces of the intellectual world, wholly separate and distinct from one another" (Locke, 1694)

Locke's influence, though indirect, can also be seen in Saussure. He outlined the specific character of language as a system of signs in which the only essential thing was the union of meaning and sound-images, comparable to "writing-systems, the signs of deaf-mutes, symbolic rites, polite formulas, military signals, etc."

From this he progressed to the idea of *semiology,* the name Saussure preferred for semiotics:

"A science that studies the life of signs within society is conceivable; it would be a part of social psychology ... I shall call it semiology. Semiology would show what constitutes signs, what laws govern them. Since the science does not yet exist, no one can say what it would be; but it has a right to existence, a place staked out in advance. Linguistics is only a part of the general science of semiology; the laws discovered by semiology will be applicable to linguistics... language, better than anything else, offers a basis for understanding the semiological problem... If we are to discover the true nature of language we must learn what it has in common with all other semiological systems. By studying rites, customs, etc. as signs, I believe that we shall throw new light on the facts and point up the need for including them in a science of semiology and explaining them by its laws" (Saussure, 1986).

Succeeding writers on semiotics have carried the argument to one or the other extreme. Kristeva sees that for Saussure linguistics is only part of a general science of signs and proposes that linguistics in due course will merge in a universal science. At the other extreme, Barthes asserts the unquestioned primacy of language from which all other sign-systems must be derived. Ever since Locke made the relation between signs (words) and ideas a central feature of human understanding, there has been debate and divergence about the nature of the sign. 'Sign' is a word used in ordinary speech; if it is to be given a more precise meaning in semiotics, it is from the ordinary meanings of 'sign' that the narrowing down must start. What then are the ordinary meanings of 'sign'?

Sign:

- A *gesture* or *motion* of the hand, head, etc. serving to convey intimation or to communicate some idea.
- A signal.
- A *mark* or *device* having some special meaning or import attached to it, or serving to distinguish the thing on which it is put.
- A *conventional mark, device or symbol*, used technically (as in music, mathematics, botany) in place of words or names written in ordinary letters.
- A *cognizance or badge.*
- Something displayed as an *emblem* or a *token.*
- A *characteristic device* attached to, or placed in front of an inn or shop as a means of distinguishing it from others or directing attention to it.
- A token or indication (visible or otherwise) of some fact, quality etc, also specifically in Medicine = *symptom.*
- The *trail or trace* of wild animals, etc.
- A trace or indication of something; a *vestige.*
- An act of a *miraculous* nature.
- One or other of the twelve equal divisions of the *Zodiac.*
- A mere *semblance* of something.
- An indication of some coming event; an *omen* or portent.

(Simpson, 1989; italics are mine)

All these possible meanings and uses of the term 'sign' are related to different interactions between the bio-psycho-social entity called 'man' and its reality.

The range of meanings and applications of the word 'sign' shows that one cannot, in semiotics, use a word as though what it means is self-evident. Peirce spent a lifetime in defining and categorising signs. At varying times he produced definitions and illustrations of 'sign' that were not all consistent with one another.

It seems, anyway, that there are a few properties peculiar and necessary to classify something as a sign (Eco, 1976):

1. It substitutes something, either by itself or by the effect it produces;
2. The substituted thing can be absent, or even not real;

3. The sign itself can be substituted by another sign;
4. The substitution can generate deceptions, because language is arbitrary and autonomous;
5. The conditions above can be satisfied in various degrees by different kinds of signs.

2.1.1 The power of negative

The ancient motto "the Word kills the Thing", expresses the negative power of thinking: Saussure extended this approach. According to Saussure, language is a system of signs in which words (the signifier) have only some *arbitrary* relation to what they signify (the actual meaning of the word). A cat, for example, has no link with the sound of the word 'cat'. "A word, therefore, gets its meaning from the way different signs relate to each other" - just as, to use one of Saussure own analogies, the meaning of a move in chess only arises out of its context within a game. For Saussure, what defines a word is not its relation to some eternal essence, but to its relationship with other words in the language system.

These relationships with other words are negative, not positive: the word does not have an absolute, essential relationship with the thing it describes; the relationship is rather purely arbitrary. The thing we call a 'cat' we could just as easily call a 'dog' or whatever. Each word (like 'bat' for example) is defined by what it is not (i.e., a 'bat' is not, is opposed to, 'bit' or 'cat' or 'bought' or 'but').

Saussure ideas set the groundwork for modern linguistics, structuralism, and semiotics, and have been pursued by later linguists such as Claude Levi-Strauss and Noam Chomsky. It is, also, at the core of the combinations and permutations which constitute information theory. In the Saussure definition of the *arbitrary* relations between signifier and signified, and between the various terms forming both these subsets, is clear that a language is *a system of relational values*. Signs are defined "not positively, by their content, but negatively, by their relations with other terms of the system" (Saussure, 1986). The nature of linguistic elements is, by this way, based on relations, combinations and differences. This is the core of Information Theory, which forms the background of any linguistic system: "…information is a difference that makes a difference" (Bateson, 1979).

Charles Sanders Peirce, Louis Ferdinand de Saussure

2.1.2 Expressions of life

Charles Sanders Peirce, in his correspondence with her, approved Lady Welby's dicta that "language is only the extreme form of expression" and that "life itself may be considered as *Expression*". We want to recall something from his letter of October 12, 1904:

"Any concept is a sign, of course. Ockam, Hobbes and Leibniz have sufficiently said that. But we may take a sign so broad a sense that the interpretant of it is not a thought, but an action or experience, or we may even so enlarge the meaning of sign that its interpretant is a mere quality or feeling. A *Third* is something which brings a First into relation to a Second. A sign is a sort of a Third. How shall we characterize it? Shall we say that a Sign brings a Second, its Object, into *cognitive* relation to a Third? …If we insist on *consciousness*, we must say what we mean by consciousness of an object" (Peirce, 1953).

He is referring to his catalogue of semiotic entities: "I divide signs into Icons, Indices and Symbols..." In this catalogue there were two basic connected triads to define the semiotic universe. One triad is defining typologies of signs; the other one is defining kinds of relations between the semiotic event, the reference and other signs.

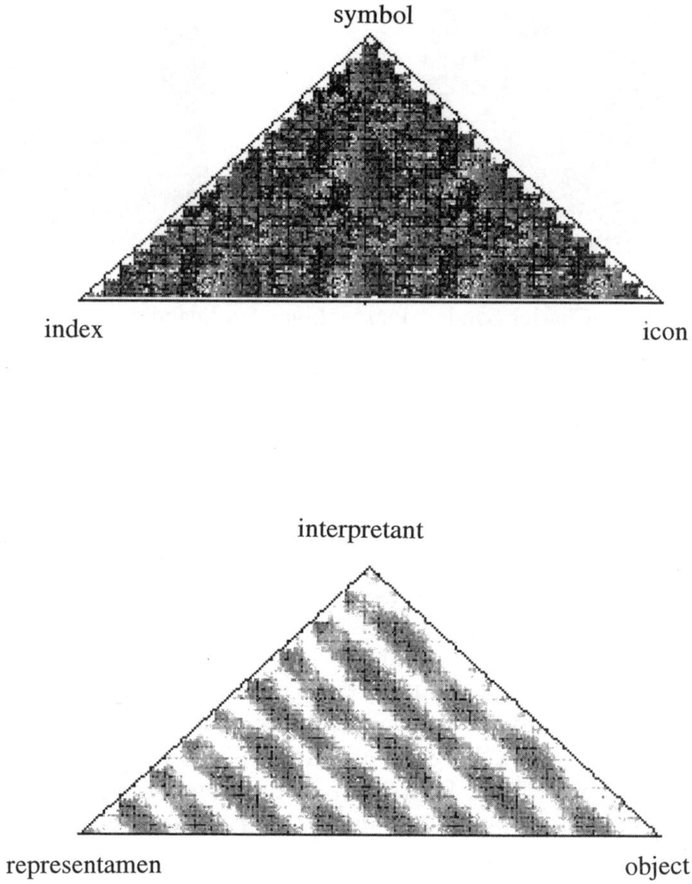

The “basic triads” in Peirce

This definition of the basic semiotic unit generates chains of relationships and geometries in the Semiotic Universe that we could simulate using an iteration of one of the two triads. The result, after 50 iterations is shown below, and it is a fractal object called “Sierpinskij carpet” if is bi-dimensional (or gasket if tri-dimensional). This can give an idea of the possible landscape of a small area of the Semiotic Universe, based on Peirce’s view (Orsucci, 1981; 2002).

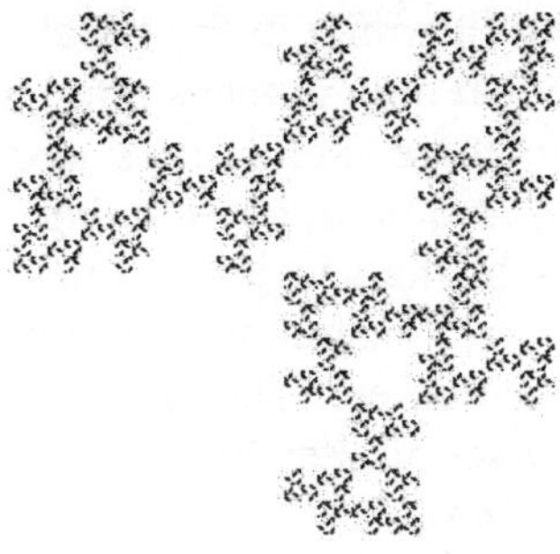 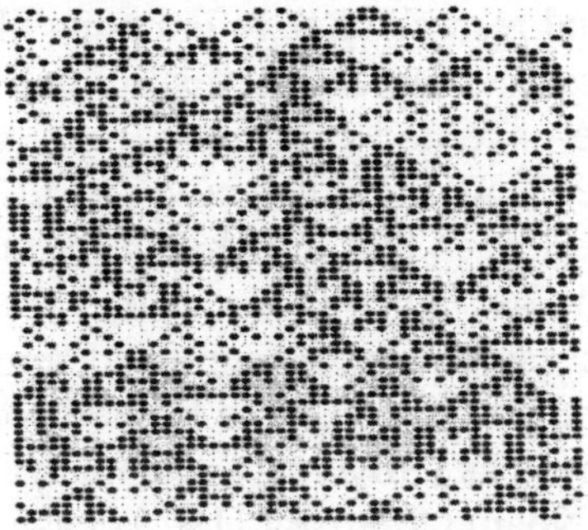

The semiotic Sierpinskij thread (left, Orsucci, 2002), and a Sierpinskij carpet, generated by a Boolean network (right, Kaplan, 1995)

The three-dimensional structure could resemble closely to a shell, but topologically made of rubber, just to be squeezed or twisted following semiotic events that change relations and distances between the triads. This simulation is consistent with the branching of meanings proposed by Ross Quillian (Minsky, 1968).

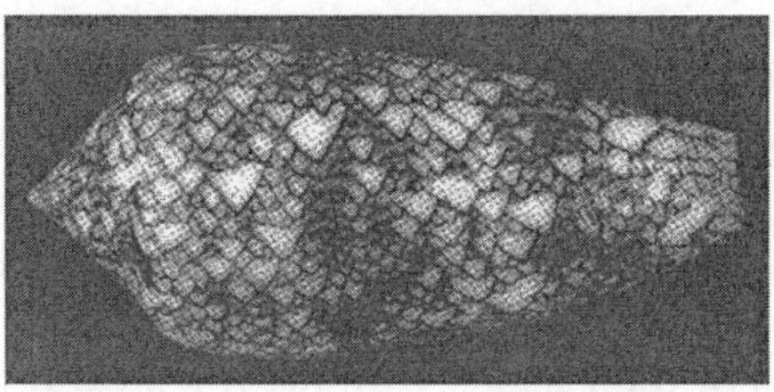

A shell as a topology for the semiotic universe (Kaplan, 1995)

A closer look at the Quillian scheme and at our simulation of the Sierpinskij thread may suggest that the triadic structure suggested by Peirce can be a little bit simplifying, but its heuristic power has still many suggestions to offer. The actual forms of the Semiotic Universe are biomorphic, as the shell design similarity shows.

It has been argued that grammars represent the class of possible solutions to the problem of mapping hyper-dimensional meanings onto a low-dimensional channel, heavily constrained by the limits of human information processing (MacWhinney & Bates, 1989). There may be a casual resemblance to domains like birdsong (learning in the vocal channel), chess (a complex set of solutions to a game that only humans play), or music (rule-governed transitions in sound) but these similarities are largely superficial. The *meaning space involved in language-mapping* includes experiences that are shared by all members of the species and the channels used by human language are subject to constraints on information processing: e. g. perception, memory, articulation etc. The class of solutions to the problem is quite limited constituting a set of alternatives that might be referred as Universal Grammar (Chomsky, 1983).

Noam Chomsky (1975) has proposed that "Linguistic theory, the theory of UG [Universal Grammar]... is an innate property of the human mind", and that we should conceive of "the growth of language as analogous to the development of a bodily organ". Indeed, Chomsky has argued that "a general learning theory....seems to me dubious, unargued, and without any empirical support (1982). Where would such rich innate structure reside?

Pinker (1994) suggests that this innate knowledge must lie in the micro-circuitry of the brain. If the notion of a language instinct means anything at all, it must refer to a claim about cortical microcircuitry, because this is the only way that detailed information can be laid out in the brain. This kind of representational nativism is theoretically plausible and attractive, but it has proven hard to defend on both mathematical and empirical grounds.

On mathematical grounds, it is difficult to understand how 10^{14} synaptic connections in the human brain could be controlled by a genome with approximately 10^{6} genes, particularly when (a) only 20-30% of these genes at most go into the construction of a nervous system, and (b) humans share approximately 98% of their genes with their nearest primate neighbours.

Paul Churchland (1995) reminds us that each synaptic connection can take multiple values. If we assume conservatively that each connection can take 10 values, Churchland calculates that the synaptic coding power of the human brain contains more potential states of connectivity than there are particles in the universe! Genes would need a lot of information to orchestrate a system of this size.

Of course a detailed mapping from genes to cortex would still be possible if genes behaved like letters in the alphabet, yielding up an indefinite set of combinations. But this is not the case; instead, genes operate within a highly constrained spatiotemporal and chemical matrix, using and reusing topological principles that have been conserved over millions of years and thousands of species. The most plausible way to think at the sources of the Universal Grammar that shapes the Semiotic Universe is to see its forms as emerging from the interactions between the genetic substratum, sensory-motor activities and the world (Bates, 1976)

2.2 Language as a dynamical system

When we study brain related activities we usually have two risks:

1) considering brain as the other body organs. Brains 'carry out computation' (it is argued); they 'entertain propositions'; and they 'support representations'. Brains may be organs, but they are very different.

2) considering brain as similar to a digital computer. The brain does some things which are very similar to human-made symbol processors; but there are also profound differences between the brain and digital symbol processors, and attempts to ignore these on grounds of simplification or abstraction run the risk of misunderstanding the nature of neural computation (Churchland, 1992).

Language processing is traditionally assumed to involve a lexicon, which is the repository of facts concerning individual words, and a set of rules which constrain the ways those words can be combined to form sentences. Although the processing of language clearly unfolds over time, the representations which are produced by traditional models typically have a curiously static quality. This is revealed in several ways. For instance, it is assumed that the lexicon pre-exists as a data structure in much the same way that a dictionary exists independently of its use. Similarly, the higher-level structures created during sentence comprehension are built up through an accretive process, and the successful product of comprehension will be a mental structure in which all the constituent parts (words, categories, relational information) are simultaneously present.

If language processing is seen as taking place in a dynamical system, the lexicon is viewed as consisting of regions of state space within that system; and the grammar consists of the dynamics (attractors and repellers) which constrain movement in that space. This approach entails representations which are highly context-sensitive, continuously varied and in which the objects of mental representation are better thought of as trajectories through mental space rather than things which are constructed (Abraham, 1990.

Higher-level categories correspond to large regions of state-space; lower-level categories correspond to more restricted sub-regions. The image of language processing just outlined does not look very much like the traditional picture which we began with. *Instead of a dictionary-like lexicon, we have a state space partitioned into various regions. Instead of symbolic rules and phrase structure trees, we have a dynamical system in which grammatical constructions are represented by trajectories through state space.* Over recent years, there has been considerable work in attempting to understand various aspects of speech and language in terms of dynamical systems. Some of the most elegant and well-developed work has focused on motor control, particularly within the domain of speech. Some of this work makes explicit reference to consequences for theories of phonology.

2.2.1 Language forms

If "to imagine a language is to imagine a form of life" (Wittgenstein, 1967) this means that language is the basic stuff of thinking, culture and the organization of perception and action. Every raw perception or action is inscribed in a semiotic texture. This is a dynamic order arising and returning to the experience of the world (Merleau-Ponty, 1945). These different orders emerge from the habits and uses we have in everyday social communication.

According to Jacobson (1966) "the Addresser sends a message to the Addressee. To be operative the message requires a *Context* ('referent' in another, somewhat ambiguous nomenclature), sizeable by the *Addressee*, and either verbal or capable of being verbalized; a *Code* fully, or at least partially, common to the *Addresser* and *Addressee* (or in other words, to the encoder and decoder of the message); and, finally, a *Contact*, a physical connection between the Addresser and Addressee, enabling both of them to enter and stay in communication."

If communication is oriented toward context, then the *referential* function dominates ("Ottawa is the capital of Canada"); if it is oriented toward the Addresser of the message, then the *emotive* function dominates ("I fall upon the thorns of life! I bleed!"); if it is oriented toward the Addressee of the message, then the *conative* function dominates ("Watch Out for that falling brick!"); if it is oriented toward Contact, the establishment of a rapport between Addresser and Addressee so that the lines of communication are kept open, then the *phatic* function dominates ("What terrible weather we've been having"); if it is oriented toward code, then the *metalingual* function dominates ("Understand what I mean?"); and, last, if it is oriented toward the message for its own sake, then the *poetic* function dominates ("The force that through the green fuse drives the flower / Drives my green age").

The important implication of Jacobson's theory is that *meaning does not reside in the message per se; it is part of the total act of communication*, not a stable entity which passes, uncontaminated, from Addresser to Addressee. A Borromean Knot is a clever way to represent the dynamical complementarities of these components. In this topological scheme three rings are bonded in a way that if you cut just one of them the other twos fall apart.

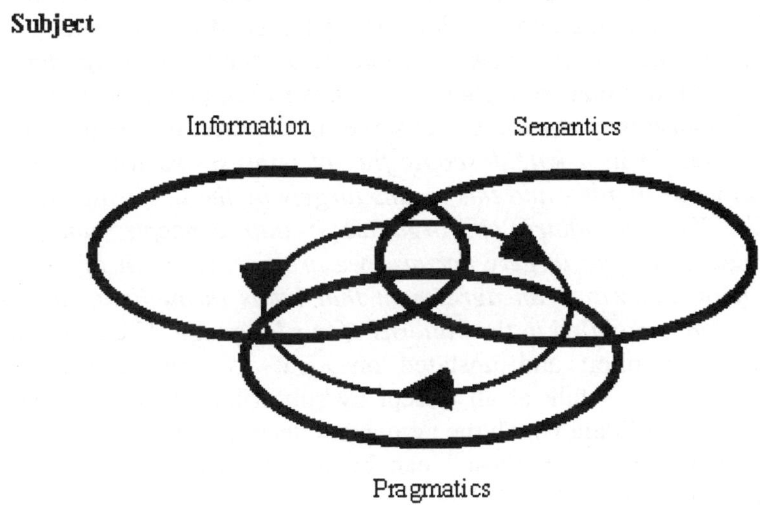

2.3 Language and reality

The Sapir-Whorf theory, named after the American linguists Edward Sapir and Benjamin Lee Whorf, is a "mould theory" of language. Writing in 1929, Sapir argued in a classic passage that:

"Human beings do not live in the objective world alone or alone in the world of social activity as ordinarily understood, but are very much at the mercy of the particular language which has become the medium of expression for their society. It is quite an illusion to imagine that one adjusts

to reality essentially without the use of language and that language is merely an incidental means of solving specific problems of communication or reflection. The fact of the matter is that *the 'real world' is to a large extent unconsciously built upon the language habits of the group. No two languages are ever sufficiently similar to be considered as representing the same social reality. The worlds in which different societies live are distinct worlds, not merely the same world with different labels attached... We see and hear and otherwise experience very largely as we do because the language habits of our community predispose certain choices of interpretation* (Sapir, 19569. This position was extended in the 1930s by his student Whorf, who, in another widely cited passage, declared that:

"*We dissect nature along lines laid down by our native languages*. The categories and types that we isolate from the world of phenomena we do not find there because they stare every observer in the face; on the contrary, the world is presented in *a kaleidoscopic flux of impressions which has to be organized by our minds - and this means largely by the linguistic systems in our minds. We cut nature up, organize it into concepts, and ascribe significances as we do, largely because we are parties to an agreement to organize it in this way - an agreement that holds throughout our speech community and is codified in the patterns of our language*. The agreement is, of course, an implicit and unstated one, but its terms are absolutely obligatory; we cannot talk at all except by subscribing to the organization and classification of data which the agreement decrees. (Whorf, 1956)

The 'Sapir-Whorf hypothesis' can be described as consisting of two associated principles:

- *linguistic determinism*, our thinking is determined by language.
- *linguistic relativity*, people who speak different languages perceive and think about the world quite differently.

There are many examples we can cite about this, the Eskimo names for snow is the most frequent.

While few linguists would accept the Sapir-Whorf hypothesis in its strong, extreme or deterministic form, many now accept a more moderate perspective, namely that the ways in which we see the world are influenced by the kind of language we use:

- the emphasis is on the potential for thinking to be influenced rather than unavoidably determined by language;
- it is a two-way process, so that the kind of language we use is also influenced by the way we see the world;
- any influence is ascribed not to Language as such or to one language compared with another, but to the use within a language of one variety rather than another (typically a *sociolect* - the language used primarily by members of a particular social group);
- emphasis is given to the social context of language use rather than to purely linguistic considerations, such as the social pressure in particular contexts to use language in one way rather than another.

The synthesis of a dynamical view of the Sapir-Whorf theory is that as language organizes and moulds sensory experiences, in the same way it is changed by action and perception. All these "phase transitions" between language and reality happen at the information (indexes) layer and interface. It is the same layer which is posited at the interface with the brain: we have to suppose a constant flux of (sub-symbolic) information between brain, mind-language, and reality

Information

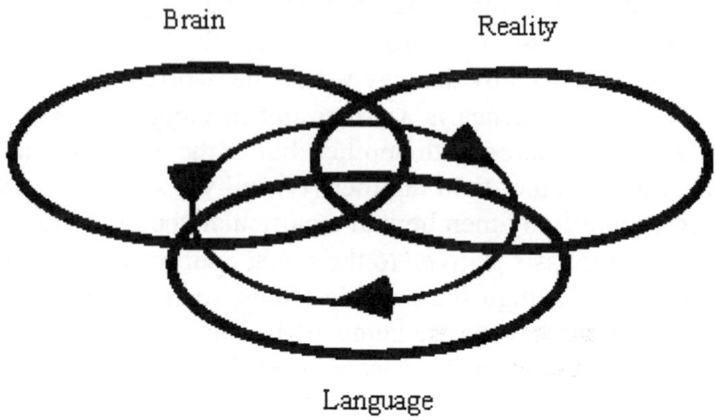

2.4 Olfaction

Smell is functioning on the borders of digital discrimination and continuous sensory perception; language and nature.

Plato refused any place in the heaven of ideas for smell: the universals for odors are quite problematic. That probably recalls that Western thinking has been built upon the primacy of sight. We can just remember that the root of the word 'idea' is the Sanskrit '*id-*' 'to see'.

Following Aristotle smell is not quite an intellectual sense. Its position in the sensory hierarchy is at the transition between distal modalities, such as vision and hearing, which mediate distant signals, and proximal modalities, such as taste and touch which suppose direct contact.

An odor is also sensory *persistent*: different odors superpose each others, creating strange cocktails. The intermediary position of olfaction may explain its low involvement in abstraction. This difficulty also bears advantages because it is at the origin of the considerable evocative, emotional, power of smells. Because olfactory sensations might be better preserved than any other sensory impression from the decomposition of intellectual analysis, they might act as idiosyncratic signs: at the transition between indexes and icons. One can thus understand, as Gaston Bachelard rightly expressed it, that "an odour is in one's childhood, in one's life ... *an immense detail*". Or remind the smell and taste of the *Madeleine cookies* in the *Recherche* by Marcel Proust.

It is difficult to find words to describe taste and smell; you have to use metaphors, analogies and all sort of poetic words used to describe food, perfumes and wines. It is interesting to have a look at the attempts to build *categories and metrics*, to build a classification of odors.

Carolus Linnaeus (1707-1778) is famous for his system for the classification of living species. It is almost unknown that he also suggested a seven category system for odors:

1. Champhoraceus
2. Musky
3. Floral
4. Pepperminty
5. Ethereal
6. Pungent
7. Putrid

Later, Hendrik Zwaardemaker (1857-1940) reconstructed this system and proposed nine olfactory categories:

1. Ethereal
2. Aromatic
3. Fragrant
4. Ambrosiac
5. Alliaceous
6. Empyreumatic
7. Hircine
8. Foul
9. Nauseous

Hans Henning proposed a "smell prism" which used six primary odors.

1. Flowery
2. Foul
3. Fruity
4. Spicy
5. Burnt
6. Resinous

Supposedly, any olfactory stimulus would occupy a position in this three-dimensional space. Here is a rendering:

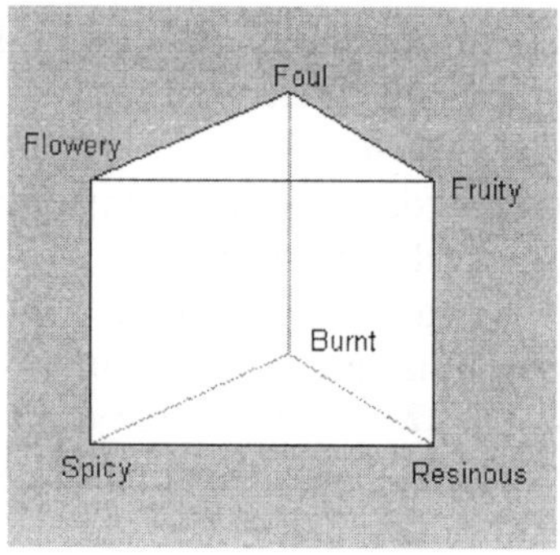

The Smell Prism (Hennig, 1896)

It is evident that *odors are indexes that generate icons, but their translation in verbal symbols is inferential and open to many different and fuzzy solutions.*

2.4.1 The smell of emotions

There is a clear example of this bio-semiotic complexity in a habit of a young girl. She used to express her emotions by bringing into her psychotherapist's office bags full of vegetables or fruits with strong smells. When she was happy she brought oranges, when she was sad she brought onions. In this example we find cultural meanings of odors, based also on their chemical effects: a smile for the sunny orange or some tears for the onion. In this way she could fill the office with the odorants she brought. She could also permeate the therapist's body by the air he breathing. Finally, she could leave a marker of her passage in the office after the end of each session, using the permanence of odors. It is evident the *heterogeneity* of these signs, their multiple and intermingled functions. In this example there is an interesting mix of neurophysiology, ethology and culture (Orsucci, 1998b; 1998a).

In mammals, olfactory sensory perception is mediated by two anatomically and functionally distinct organs: the main olfactory epithelium (Minsky, 1994) and the vomero-nasal organ (VNO). The main olfactory system recognizes most odors and ultimately sends fibres to higher sensory centers in the cortex, resulting in the conscious perception of odors and measured thoughts and behaviors. The vomero-nasal system, however, detects pheromones and transmits this olfactory information via a separate pathway of neuronal projections to the emotive brain, the amygdala and hypothalamus. Thus, the VNO pathway bypasses higher cognitive centers, resulting in innate and stereotyped behavioral and neuro-endocrine responses. Unfortunately, there is no certain proof that the VNO is still active in humans.

2.4.2 Smell dynamics

Walter Freeman wrote:

"The bulb analyzes each input pattern and then synthesizes its own message, which it transmits via axons to another part of the olfactory system, the olfactory cortex. From there, new signals are sent too many parts of the brain-not the least of which is an area called the entorhinal cortex, where the signals are combined with those from other sensory systems. The result is a meaning-laden perception, a *Gestalt* that is unique to each individual. For a dog, the recognition of the scent of a fox may carry the memory of food and expectation of a meal. For a rabbit, the same scent may arouse memories of chase and fear of attack. Such knowledge has provided a valuable starting point for more detailed study of olfaction. But it leaves two important issues unresolved. The first is the classic problem of separating foreground from background: How does the brain distinguish one scent from all others that accompany it?" (Freeman, 2000).

Freeman supposed that the information relayed to the brain by odorant exposure occurred in the form of spatial patterns of pulse activity across neurons. To explore this idea and map neuronal activities on the entorhinal cortex he designed a special matrix of electrodes arranged in a 6×10 array 0,5 mm apart. In his classical experiment, he discovered that each odorant was differentiated by the spatial pattern, and not by its waveform. The sequence of learning could influence the form, and old learning patterns were changed by the new ones. This is consistent with the basic rules of system theory: a new member changes all the set. It also recalls the *Nachtraglich* (or *après-coup* in French) principle regarding memories: old memories can be changed backwards as soon as a new one is added to the file.

Memories are not stored in single cells, or in fixed networks, but change their neural patterns following the stream of time, perception and action.

Following Freeman (1998): "A hypothesis of olfactory function is that convergent sensory input is used to update a central active state covering the last few seconds by the entorhinal-hippocampal time loop, and that motor commands emitted to elicit intentional behavior through the motor loop such as sniffing are accompanied by corollary discharges called 'pre-afferent' messages through a pre-afference loop. These messages sensitize the sensory cortices to respond selectively to the expected consequences of the intended actions of exploration. By this hypothesis, sights, sounds, and touches are predicted to influence the function of the OB in ways that are to be determined. Olfactory classification is not a passive tree search. It is an active, goal-directed exploration of a continuously varying and infinitely complex chemical environment."

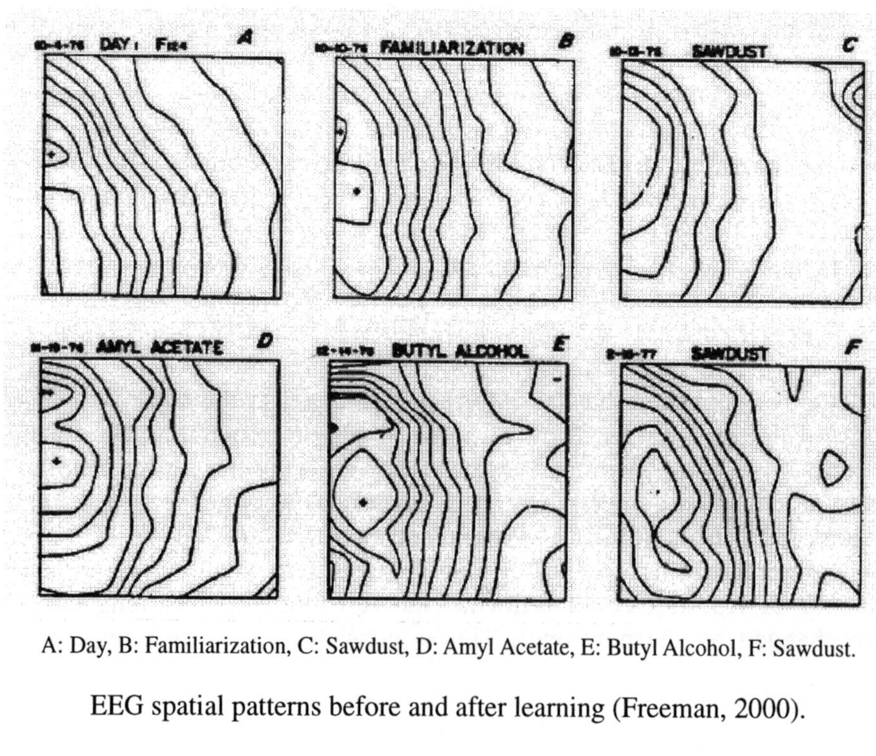

A: Day, B: Familiarization, C: Sawdust, D: Amyl Acetate, E: Butyl Alcohol, F: Sawdust.

EEG spatial patterns before and after learning (Freeman, 2000).

The information density is spatially uniform over the array. Every neuron participates in the discrimination of every odorant, irrespectively of whether it fires rapidly, slowly, or not at all, because *in pattern formation non-responding is as important as responding*. It is also fully compatible with pattern recognition theory, in which an integral transform of an image, such as the 2-dimensional Fourier transform, serves to de-localize the features of patterns. This is how a hologram, for example, can be broken into pieces, but each piece has the full image, though with reduced resolution. "These results show that the odorant-related patterns are not fixed representations of odors, nor can they be stored, retrieved or matched as such. They are dynamical action patterns" (Freeman, 1998).

We remember that selective amplification is set by *Hebbian learning and habituation: recalled synapses are reinforced by rewards*. Responses to odours in the OB take the form not of retrieval, but of fresh construction by nonlinear dynamics. The constructs are shaped by synaptic modifications from past experience, and by pre-afferent signals from the limbic system. Synapses in the olfactory system are modified during learning, not to store memory traces of odours that are to be recalled and matched with new odorant patterns, but to guide the dynamic construction of OB activity patterns as a basis for classification following identification. It is important to be aware that these structures reflect the meaning of the stimulus, not merely its presence. *The spatial structure of 'peppermint' associated with positive reinforcement, for example, is different from the structure for the same odorant but associated with punishment.*

2.4.3 The epistemology of olfaction

Visual metaphors pervade our everyday discourse about cognition (e.g. in French, voir/savoir, in English see/understand). Indeed, it seems quite natural to us to equate thinking and seeing. Is this association a human universal? The anthropological record suggests otherwise. The Surya Indians of Brazil equate hearing and understanding; the Ongee people of the Andaman Islands - in a case that is of particular interest to us - treat smell as the medium for thought. In some sense, they built an entire epistemology upon the primacy of smell. These examples suggest that the link between vision and knowledge (which appears so natural to us) is a matter of convention, rather than of nature. It also suggests exploring all the cross-modal, cross-sensory, activities.

What are the cognitive consequences of a culture selecting some other sense than vision as its dominant lead for cognition? Do the Ongee, for example, reckon space in the same way Westerners do? How different is a smell space from a visualized one? Does the ostensible transparency of our own visual model of mind perhaps occult certain aspects of cognition from our view?

2.4.4 A model for olfaction

This model may also account for the integrative neurobiology of vision, audition, and somesthesia, particularly the binding problem, for three reasons:

1. in the evolution of vertebrate forebrains olfaction preceded all other senses, and set the direction of development of algorithms for perception;
2. all sensory systems converge to the Entorhinal Cortex, and they must develop and share a common mode of function in order to support the emergence of multi-sensory perceptions;
3. the spatiotemporal patterns of EEGs from the sensory neocortices show the same changes with learning and have the same spectral and statistical properties as the olfactory EEGs.

A flow diagram is constructed from anatomical connection data and used to specify the inputs and outputs of a set of coupled Ordinary Differential Equations. Each node, Fi, sums inputs from all other nodes, vi, after the sigmoid conversion, G, and the synaptic weights, k_{ij}, $i \neq j$ (no neuron acts synaptically onto itself):

$$F_i = \Sigma\, k_{ij}\, G\,[v_j] + n_j$$

where n_j is input, including Gaussian noise that is required to stabilize chaotic attractors in the model and to simulate receptor input. The Olfactory Bulb layer is modelled by 64 oscillators, each containing interactive M sets and G sets, and its own P set. The L1 to L4 operators model axonal tract dispersion and delay with linear 2nd order low pass filters. This constitutes the KIII model of the olfactory system. From Yao et al. (1990) in Freeman (2000).

2.5 Neural learning

Freeman suggests the existence group of neurons called *nerve cell assemblies*, which are stimulated by other neurons during learning. The behavior of these cells is dictated by the Hebb's Rule: *synapses between neurons that fire together become stronger, as long as the synchronous firing is accompanied by a reward.* These neurons effectively connect the sensory input neurons to each other by receiving input from several of them at once. When simultaneous excitation among them occurs in connection with a consequence, the Hebbian synapses of the nerve cell assembly are selectively strengthened. That strengthening is reflected in increased sensitivity in the post-synaptic cell. The consequence of such a system is that it would help coordinate the burst which occurs with recognition, and that this would then help the input associated with a familiar scent stand out against less important incoming information. These assemblies thus strengthened would be physical "repository for past associations", reflecting not just recognition of a smell but the experiences which went along with it, assuming that the experiences were significant in some way.

The second influence which modulates our sensitivity to familiar scents is not hypothetical, but would complement the activity of the postulated nerve cell assemblies. This is the *priming* of the Olfactory Bulb, which has the overall effect of increasing sensitivity. One way this is accomplished is through system-wide arousal, like hunger or thirst, sexual attraction, or fear. These states are directed by other parts of the brain through both axonal connections to the bulb and the release of modulatory chemicals. Another way that priming is accomplished is through input itself. The Hebbian nerve cell assemblies, embedded in the primed olfactory bulb, magnify and amplify the excitatory signals of a familiar smell, so that "the input rapidly ignites an explosion of collective activity".

2.5.1 Synchrony

These are all key examples of how *the act of perceiving is driven as much by internal state and past experience as it is by the object perceived: it is as subjective as it is objective.* And, at least in part, the sensory system is capable of modulating itself and its recognition and its response - it is not just reporting, it is interpreting as well.

The synchronized activity of the neurons in the Olfactory Bulb upon recognizing an odor has the effect of putting the message in capital letters and then putting several exclamation points on the end. The *synchronized signals are additive,* whereas the unsynchronized signals reflecting irregular everyday life essentially cancel each other out. This is comparable to the effect of several people in a room. If you walk into a room of twenty people who are all talking amongst themselves, you hear a murmur of voices in which individual sentences or stories are hard to distinguish. If, however, those twenty people are all singing the same song, then you get the message loud and clear. Such are the effects of *synchrony.* Thus, a synchronized signal has priority: it insures that the olfactory cortex gets the message. This works because of the extensive branching that links the two areas, so that each bulb neuron synapses with several thousand cortical cells, and each cortical cell receives input from several thousand bulbar cells.

Presumably, the cortex has a system in place for generating its own collective bursts, which one would expect to be qualitatively different from those of the bulb. If this is so, it would indeed act as a filter, screening out the ambient noise of unfamiliar smells and transmitting only the news of something already known. This collective burst would then be transmitted to other parts of the brain, a message with its edges smoothed out and its salient features highlighted. Primarily, these messages seem to go to the entorhinal cortex and the limbic system, "where signals are combined with those from other sensory systems" (Freeman, 2000).

The system of cross-wiring between bulb and cortex has another important consequence: it allows any cell in the cortex to "view" the activity of several thousand bulbar cells. Thus, in the one direction, the message from the outside world is refined, while in the other direction, the window through which one part of the brain views the world is enlarged.

2.5.2 Memory of smell

It leads us, finally, to a model of the relationship between memory and smell, and to an idea of how *Gestalten* might be formed. It relies rather heavily on this progression through layers, in which every step is a kind of distillation of the message: bulk is lost, but clarity and concentration is gained. Thus, if there are ways for the cells of the limbic system to build associative relationships in the same way that the olfactory bulb can, and they are receiving messages from all five senses, and the associations are reinforced by experience, then we have described an organization which links *multimedia* perspectives, and assigns meaning to them.

It provides a possible basis for understanding why, for example, a smell which reminds us of something often accompanies a heightened sense of awareness: the *smell is part of a cascade*, and it has alerted our bodies to look out for the other, corroborating parts. Perhaps the act of remembering is just that: a way for our senses to look around for something that used to be there, and to take stock of what changes have occurred. *The fact that the limbic system is involved in both memory and emotions certainly provides some clue as to why the act of remembrance can be so emotionally laden.*

Another satisfying aspect of this admittedly enormously incomplete model is that it accounts for changing perspective. Memories don't go away as much as they morph over time, with new experiences, shifting to accommodate new patterns but never completely disappearing. Although the experiments described were done with animals in a specific, rewarded learning system, it is not hard to imagine similar patterns guiding more complex experiences. It is not quite so important to know precisely how such processes differ in a non-controlled environment, or for that matter in a nervous system which has a "more developed" anatomy, or *sense of self.*

2.5.3 Orders

We examined clinical examples and empirical studies which suggested the importance of sub symbolic and often "hidden" interactions. A better definition of their presence and organization could be important. A possible approach could lay in studies on sensorial and perceptual interaction, while another important approach resides in studies on sub-symbolic patterns in language and communication.

We could investigate some forms in language, its morphology. This term was originally used in biology, but since the mid of this century has been used to describe the type of investigation which analyzes all those basic 'elements' which are used in language. This kind of 'elements' are technically known as morphemes. At this level of the semiotic universe we can study the forms built by different combinations of basic elements.

2.6 Human information

Five decades ago Shannon (1949) established the fundamental concept of *entropy for human writings*, which is a meaningful measure of information content In a general sense, a novel, a poem, a play of music, a speech transcription, a computer program, a DNA or protein sequence, etc. can be regarded as a "n- dimensional string of elements". Grassberger has devised a scheme to estimate Shannon's entropy for very long strings. The most commonly used method to probe the correlation in a string of elements has been the Fourier power spectrum, but recently other methods have been proposed. A new measure of complexity which is a generalization of entropy was introduced by Y.C. Zhang as the exponent α which is related to the usual power spectrum exponent β by the following relation:

$$\alpha = \frac{\beta + 1}{2}$$

where β is defined by the power spectrum. The value of 1/2 α implies that the string is not long range correlated, and it leads to white noise, while α = 1 corresponds to *the scale invariant 1/f noise*, which is also called the *maximum complexity limit*.

Along with this kind of approach Schenkel analyzed several long texts from H.G. Wells's novels, to computer programs, to the Bible and Koran (in various translations). They found several grades of complexity ranging from a $\cong 0.52$ ("The Time Machine") to a $\cong 0.986$ (a computer program), and concluded that "one may wonder what is behind the particular shape of a writing". They reached this conclusion by a preliminary transformation of letters in binary numbers which can generate a "random walk model".

M. Amit following an analogous approach studied the long range correlations in various translations of the Bible (as a sample text) by mapping them onto a one dimensional random walk model. They concluded that the experiment varies from one language to another, while it takes its maximal value in the original version i.e. Hebrew for the Bible: this means that *translations reduce complexity*. They assert in addition, that correlation may be due to two reasons: *correlation between ideas, and long range correlation due to syntax structure*.

They suggest changing name to the α value which could be better defined as *rigidity*, as the more the syntactic structure is rigid (as in computer programs) the higher are the values. They finally suggest that hidden coded (encrypted) information can be embedded in texts as has been shown in studies on equidistant letter sequencing (ELS) on the Hebrew Bible.

A phase-space approach has been applied to the analysis of transcriptions from speech of psychiatric patients. This analysis implied the construction of the phase space plot, and measurements of the Hausdorff dimension and Lyapunov exponents. This method highlighted some interesting differences between pathologies: for example, dimensions for depressed patients were different from schizophrenics and normal controls (Orsucci, 1996)

2.6.1 Kinds of orders

Language can be considered to consist of the following layers for which abstract descriptions are self-similar.

Written language:

1. letters make up words in a structured way,
2. words make up clauses in a structured way,
3. clauses make up sentences in a structured way,
4. sentences make up paragraphs in a structured way,
5. paragraphs make up chapters in a structured way,
6. chapters make up books in a structured way, etc.

Spoken language has a similar make up but it is based on wave forms rather than letters. The dimensionality of speech wave forms is 1.66. This seems to be a result of the continuous nature of the wave forms as opposed to the digital nature of the written language.

At the word level G. K. Zipf (1949; 1965) described a number of common properties of natural languages including two "laws" of word frequencies. He proposed that these laws were the result of his controversial *Principle of Least Effort*. The two laws of word frequencies are:

1. The Rank-Frequency law.

This is the most famous one; unfortunately many people call it "Zipf's law" as if it was the only one. The procedure to estimate this relation is very simple: the tokens in a text are sorted by decreasing frequency and a rank number is assigned to each token. For tokens with the same frequency, the subsorting and ranking is arbitrary. The plot of *log(frequency)* [y axis] *versus log(rank)* [x axis] approximates a straight line of slope -1.

2. The Number-Frequency law.

Being n the frequency of a token, the plot of *log(n)* [y axis] *versus log(number of tokens with frequency n)* [x axis] approximates a straight line of slope -0.5. While the rank-frequency law tends to occur for the high frequency tokens (although not necessarily for the first few ranking positions), the number-frequency law is observed for the low frequency tokens.

There have been a number of reports confirming Zipf's laws (almost exclusively the rank-frequency one) in a large number of languages, from alphabetic (most European languages), to syllabic (Japanese) and logographic languages (Kristeva, 1986). To complicate things further, random texts also show a tendency to follow the rank-frequency law, although the reasons for this are well understood and different than for natural languages. There are suggestions that these laws have some relation with the way we process and store information and therefore apply to all languages despite their origin.

To account for broad scale distributions in space and time in interacting dynamical systems, Bak, introduced the notion of *Self-Organized Criticality (SOC) to explain 1/f phenomena like the Zipf laws* (Bak, 1996).

2.7 Combinations

At the *orthographic* level Claude Shannon (1949) defined other orders when he stated: "In the teletype case where all symbols are of the same duration, and any sequence of the 32 symbols is allowed the answer is easy. Each symbol represents five bits of information. If the system transmits n symbols per second it is natural to say that the channel has a capacity of 5n bits per second. This does not mean that the teletype channel will always be transmitting information at this rate - this is the maximum possible rate and whether or not the actual rate reaches this maximum depends on the source of information which feeds the channel, as will appear later." *Orthography, being the art of writing words with the proper letters according to standard usage, is the representation of the sounds of a language by written or printed symbols. In this way, orthography is a way to access the combinatorial mathematics of information theory, at the sub-symbolic level of natural language.*

An impressive demonstration of the unexpected neural effects of orthographical combinatorics is given, for example, by a research on "the cultural effect on brain function" (Paulesu E & et al., 2000). They presented behavioral and anatomical evidence for a multi-component reading system in which different components are differentially weighted depending on culture-specific demands of orthography. Italian orthography, for example, is consistent, enabling reliable conversion of graphemes to phonemes to yield correct pronunciation of the word. English orthography is less consistent, and has complicated mapping of letters to word sounds. In this behavioral study, Italian students showed faster word and non-word reading than English students. In two PET studies, Italians showed greater activation in left superior temporal regions associated with phoneme processing. In contrast, English readers showed greater activations, particularly for non-words, in left posterior inferior temporal gyrus and anterior inferior frontal gyrus, areas associated with word retrieval during both reading and naming tasks. English readers had to make contextual associations to find the exact pronunciation of written words, while in Latin languages the connection between how you read and how you pronounce is more direct, and doesn't need activation of accessory brain areas.

This is an interesting, simple and elegant, way to have some clues about what happens at the transitional areas between mind, brain and environment.

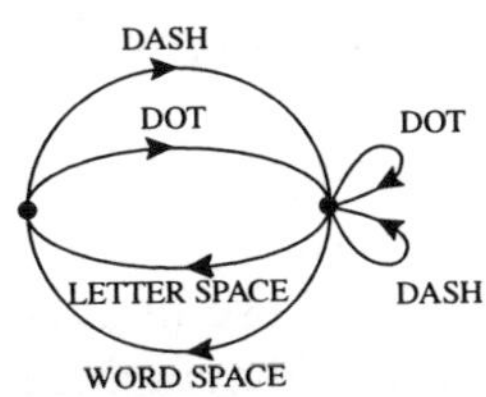

Graphical representation of the constraints on telegraph (Morse) symbols (Shannon, 1949)

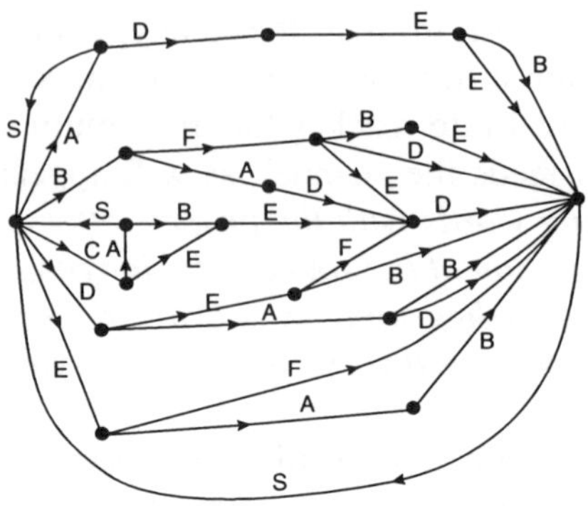

Graphical representation of the constraints
on symbols of a six letters artificial alphabet (Shannon, 1949).

2.7.1 Recurrences

Time series in natural language have also been analyzed by implementation of another methodology. This method was first introduced in the physics by Eckman, Kamphorst and Ruelle and widely used in biomedicine by Webber, Zbilut and Giuliani (see Orsucci, 1996A; Giuliani, 2002; Zbilut, 2002). Reliable metrics are needed, in order to evaluate the presence of sub-symbolic patterns and compare the different texts originated by patients' speeches. These approaches are very promising to study morphological patterns in language.

The general framework of these latter studies is originated by the idea that each subject has a personal blueprint in speech which, in semiotics, is called *idiolect*. The idiolect represents the unique structure of the individual's linguistic system.

Human relations could be defined as interactions and couplings between idiolects, as they happen on multiple levels: symbolic and sub symbolic (icons, indexes- information).

In this direction the work by Nicolis & Katsikas (in West, BJ 1996) on chaotic dynamics of linguistic-like processes, at the syntactical and semantic levels, is an attempt to mode1 these linguistic complexities, in the pursuit of a multi-fractal attractor. The orthographical level of language and thinking can be posited at the interface with the neurobiological structures. This is the level in which the bidirectional transduction processes of information between mind and body is always active. Couplings of informational attractors can drive the body-mind dynamics, as well as been implied in the interpersonal and environmental interactions.

A methodology based upon Recurrence Quantification Analysis has been proposed for the study of orthographic structure of written texts. A recurrence plot is a 2-dimensional N x N pattern of points where N is the number of embedding vectors obtained from the delay coordinates of the input sequence. From the occurrence of lines parallel to the diagonal in the recurrence plot it can be seen how fast neighboured trajectories diverge in phase space. Therefore, the average length of these lines is a measure of the reciprocal of the largest positive Lyapunov exponent.

Recurrence plots help revealing phase transitions and non-stationarities. Visible rectangular block structures with a higher density of points in the recurrence plot indicate phase transitions within the signal.

Parameters are:

1. number M of neighbors
2. step size s (i.e. size-1 samples are skipped between successive columns of the plot)
3. embedding dimension m
4. relative radius r

RQA main parameters are:

- **#recurs** = Total number of recurrent points in upper triangular area.
- **%recur** = Percent recurrence = **#recurs** / triangular area.
- **%determ** = Percent determinism = #recurrent points forming upward diagonal lines / **#recurs** (set equal to –1 if **#recurs** = 0)

Five different orthographic data sets (20th century Italian poems, 20th century American poems, contemporary Swedish poems with their corresponding Italian translations, Italian speech samples, and American speech samples) were subjected to recurrence quantification analysis.

From a strictly mathematical view, the results show the possibility of demonstrating invariance between different language exemplars, despite the apparent low-level of coding (orthography). Comparison with the actual texts confirms the ability of the method to *reveal recurrent structures, and their complexity*. Using poems as a reference standard for judging speech complexity, the technique exhibits language independence, order dependence and freedom from pure statistical characteristics of studied sequences, as well as consistency with easily identifiable texts. Such studies provide phenomenological markers of hidden structure as coded by the purely orthographic level.

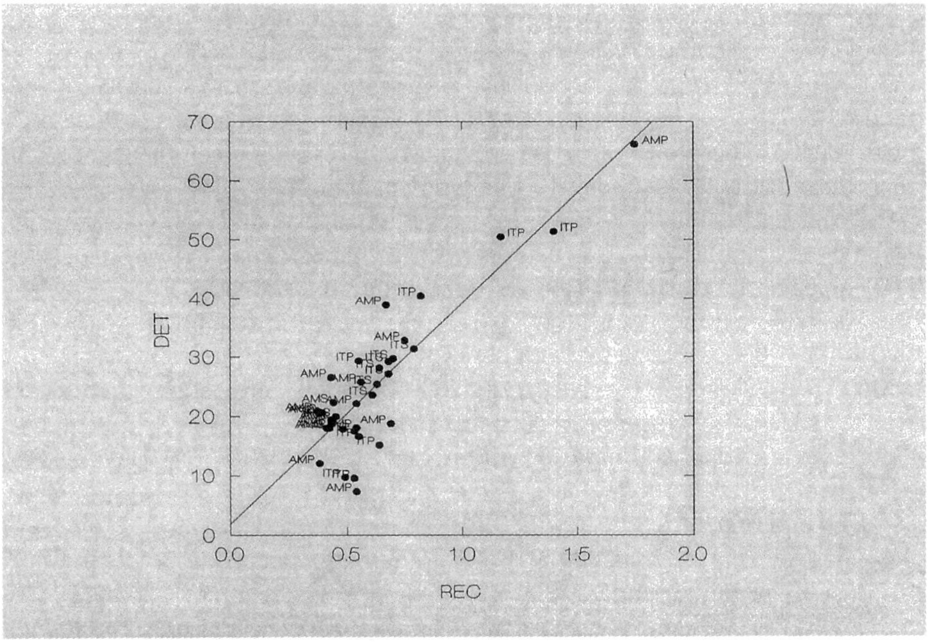

Distribution of the RQA results of different orthographic samples (recurrence vs. determinism) (Orsucci et al, 1996).

This informational level reflects orders on the word and the phrase levels. This kind of experimental approach makes possible new explorations on linguistic and cognitive structural coupling in inter-subjective dynamics.

2.7.2 The Voynich mystery

The *Voynich Manuscript*, which has been dubbed "the Most Mysterious Manuscript in the World", is named after its discoverer, the American antique book dealer and collector, Wilfrid M. Voynich, who discovered it in

1912, amongst a collection of ancient manuscripts kept in villa Mondragone in Frascati, near Rome, which had been by then turned into a Jesuit College (closed in 1953). From a piece of paper which was once attached to the Voynich manuscript, and which is now stored in one of the boxes belonging with the Voynich manuscript holdings of the Beinecke library, it is known that the manuscript once formed part of the private library of Petrus Beckx S.J., 22nd general of the Society of Jesus. The manuscript counted at least 116 folios, of which 104 remain. The folio size is 6 by 9 inches, but some folios are two or three times that size and are folded. There is one large composite of six times this size (18 by 18 inches). Both the illustrations and the script of the manuscript are unique. As long as the script cannot be read, the illustrations are the only clue about the nature of the book. According to these illustrations, the manuscript would appear to be a scientific book, mostly an illustrated herbal with some additional sections.

Wilfrid Voynich judged the Manuscript to date from the late 13th century, on the evidence of the calligraphy, the drawings, the vellum, and the pigments. It is some 200 pages long, written in an unknown script of which there is no known other instance in the world. It is abundantly illustrated with awkward coloured drawings. Drawings of unidentified plants; of what seems to be herbal recipes; of tiny naked women frolicking in bathtubs connected by intricate plumbing looking more like anatomical parts than hydraulic contraptions; of mysterious charts in which some have seem astronomical objects seen through a telescope, some live cells seen through a microscope; of charts into which you may see a strange calendar of zodiacal signs, populated by tiny naked people in rubbish bins.
Despite a number of supposed solutions to the mystery of the *Voynich Manuscript*, today it remains unread. It is unknown in which language it was written and whether it is in a code (cipher) or in plain text, using abbreviations and the alphabet is unknown. There is also the possibility of a hoax, so at this stage one may well ask whether there is an underlying language at all. Plain text, code, gibberish? Here are reported a number of features found in the manuscript that are characteristic of natural languages. Perhaps the results shown here pose no definite proof of an underlying language in the manuscript, mainly for two reasons:

- it is unknown what a character in the manuscript represents (a character, a syllable, and a prefix/suffix, a mixture of all these or just a code?);

- it is unknown whether the space between two words is in effect a word separator or a null character.

Nevertheless, some features suggest that we are in the presence of a natural language, even if it is not known which one. In this case, the assumptions are:

- the space indicates a word separator;
- the script represents either characters, syllables or abbreviations (meaning that the same words are encoded always in the same manner);
- a token is therefore any string of characters separated by spaces;
- a word is a type of token regardless of its frequency in the text.

Based on these conjectures Gabriel Landini (unpublished) investigated the Zipf's laws, and found that the manuscript follows these rules, in this way resembling written in some, unknown, natural language. In the case of the Voynich manuscript the 1/f trend suggests that there is a complex order to be explored.

The trickery of the Manuscript is that it is mediating some form of unknown knowledge in an unknown way. San Juan de la Cruz used to say: "To reach the place you don't know you have to take the path you don't know".

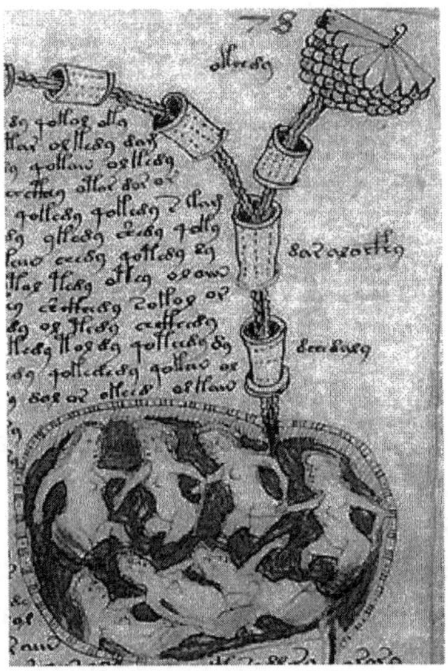

Page from the Voynich Manuscript

2.8 Mediations

According to Ferdinand de Saussure (1986), semiotics (or semiology as he called it) was to study "the life of signs in society"; and the second mythical founding-father, Charles Sanders Peirce, as well as his forerunner John Locke, conceived of semiotics as being the "doctrine of signs". Later in life, however, Peirce came to prefer the wider term *mediation* as a description of the subject matter of semiotics. And Saussure actually argued that in the semiotic sciences, there was *no object to be studied except for the point of view which we adopt on other objects*. In a similar fashion, Umberto Eco (Eco, 1976) substituted the notion of *sign process* for the traditional *sign* concept.

It seems that the domain semiotics is the landscape of relations, where processes, dynamics are in place and transitions are usual. The sign, then, is not comprehensive enough to delimit the field of semiotics: rather, the domain of semiotics is the area of *mean-ing,* or *mediation*, in wider sense.

Chapter 3: Icon

Mind is extended:
knows nothing about it.
Sigmund Freud

3.1 Icon power

There has always been wonder about the power of forming a mental image of something not present to the senses or never before perceived in reality. Imagination has been also equated to the creative ability, or even to the ability to confront and deal with a problem. In this sense it has been recognized as resourcefulness. In philosophy has been sometimes equated with thinking or active mind. In popular culture we talk about "stories that fired the imagination". In Romantic aesthetics imagination is a creation of the mind; especially an idealized or poetic creation. But, it has also been discredited by positivism and behavioral psychology as fanciful or empty assumption.

Plato speaks, metaphorically, of an inner artist painting pictures in the soul (Plato, 1921), and suggests that memory might be analogous to a block of wax into which our perceptions and thoughts stamp impressions (*Thaetetus*). Aristotle endorses this wax impression model of memory, and describes this impression as a sort of picture (Aristotle, 1941). He introduced the notion of a mental faculty of imagination, allied to perception, and responsible for producing and recalling imagery (*De Anima*). Aristotle gave imagery a central role in cognition, he asserts that "The soul never thinks without a mental image" (*De Anima*). He maintains that the representational power of language is derived from imagery, spoken words being the symbols of the inner images (*De Interpretatione*; *De Anima*).

This was almost universally accepted in the philosophical tradition, even by non-Aristotelians, up until the 20th century. With certain qualifications and exceptions (most significantly the *clear and distinct ideas* of Descartes' epistemology), the *ideas* that played such a large role in philosophy and cognitive theory from the 17th through the 19th century are direct descendants of Aristotle's *images*. The Sanskrit root of the word *idea* is *id* which stands for 'to see' (...with mind's eye). Hume, for example, explicitly identifies ideas as images (as does even Descartes in psycho-physiological contexts).

In recent years there has been a renewed interest in the study of iconic semiotics, and imagination is a part of it. Some trends of research that start in the Aristotle's De Anima are connected even to the way our computers' interface is designed. When, at the Xerox Research Centre in Palo Alto, the icon computer interface (also called 'object-oriented') was designed, there was a long history of research on the cognitive power of icons in the background. The fact that this Human Interface Technology won its battle against other interfaces based on verbal language (as UNIX and Dos) is a stimulus for reflections.

3.1.1 Phantasma

According to Aristotle: "The soul never thinks without a mental image [*phantasma*]" (*De Anima*). In this tradition, imagery, and thus imagination, has an essential role to play in all forms of thinking (with the possible exception of direct intuitions of Platonic forms, or of the divine). And Aristotle adds: "They frequently find themselves engaged in something else besides the dream, viz. in setting a phantasm which they envisage into its mnemonic position" (*On Dreams*).

It does, however, have a special connection with *desire*. Aristotle argues that our desire for (and, thus, pursuit of) anything not actually present to the senses must be mediated by an image of the desired object. Aristotle's treatment is morally neutral, but his notion of desirous imagination may later have become conflated with the Hebraic concept of *yetser*, the wilful (but also semi-divine) faculty in humanity that led to Adam's sin (and, indeed, all subsequent). At any rate, in the Judaeo-Christian intellectual tradition (from ancient to relatively recent times) imagination, although recognized as indispensable to cognition, was usually profoundly distrusted. Unless strictly disciplined by reason it would soon lead us into concupiscence and sin.

But, of course, the connection between imagination and perception is the more fundamental. Aristotle's conception of *phantasia*/imagination seems to be closely bound up with his postulation of what came to be called the *common sense* or *sensus communis*. This is the part of the psyche responsible for the *binding* of the deliverances of the individual sense organs into a coherent and intelligible representation, and for apprehending the so called *common sensibles*, those aspects of the world (broadly, the spatio-temporal aspects) that can be known through more than one sense mode without being the characteristic proper object of any of them (Aristotelian common sensibles are broadly coextensive with Lockean *primary qualities*). In fact, it is plausible to interpret *phantasia* and *sensus communis* as different aspects or modes of a single faculty, depending on whether it is regarded as *receptive* or *productive*, or on whether it is operating in the *presence* or the *absence* of whatever is being mentally represented.

Phantasmata are generated in either case but, when their immediate cause is an object directly before us, the tendency is to refer to them as percepts, and to the process as *perception*. When memory of previously observed things is the source, reference will more likely be to imagery and *imagination*. Thus imagination came to be particularly associated with thinking about things that are not actually currently present to the senses: things that are not really there.

Some of Aristotle's successors tended to lay the stress on the conceptual separation of the notions of imagination and *sensus communis*. Thus Early Christian and Medieval anatomists often located *sensus communis* at the front of the brain's first ventricle, ready to receive sense impressions, whereas imagination was placed at the rear of this ventricle (or even in the second ventricle), and was responsible for holding and perhaps consolidating the resultant images, and passing them back to the other ventricles and faculties (where they could be used in thought or stored in memory). Imagination might also, sometimes, be held responsible for the *recombining* of various image parts into chimerical forms: the various versions of the Sphinx and her family.

3.1.2 Iconic

Saint Thomas Aquinas motivation is the effort to bring Aristotelian doctrine into conformance with Christianity. The basic Thomist premise (Thomas, 1962) is the unity and inviolability of the self that is inherent in the soul, brain and body. This unity does not allow the entry of forms (information) into the self. The impact of the world onto the senses gives rise to states of activity he called *phantasmata*, which are ephemeral and unique to each impact and therefore cannot be known. The function of the brain is to exercise the faculty of the imagination, in order to abstract and generalize over the phantasms that are triggered by unique events. These processes of abstraction and generalization create information that assimilates the body and brain to the world. *Assimilation* is not adaptation by passive information processing, nor is it an accumulation of representations by resonances. It is the shaping of the self to bring it into optimal interaction with desired aspects of the world. *The goal of an action is a state of competence* that Maurice (Merleau-Ponty, 1945) called *maximum grip*. *Assimilation* is the beginning for all knowledge, and it is rooted in *iconic thinking*.

Other ancient and medieval authors, however, were more inclined to stress the underlying identity of the imagination and the common sense, and this interpretation probably became more widespread with the increased availability of more accurate Aristotelian texts and translations in the later middle ages and renaissance. We find this tendency even in such a consciously anti-Aristotelian figure as Descartes, who, in the 'Treatise on Man', explicitly identifies both the *sensus communis* and the imagination with the surface of the pineal gland, upon which images (*ideés*) both of sense and of memory or fancy are inscribed. Although it is important not to confuse these *corporeal ideas* with the *clear and distinct ideas* that play so prominent a part in Descartes' epistemology, it is surely not coincidental that the site of the Cartesian *imagination/sensus communis*, the pineal surface, is also the privileged site of mind-body interaction, the lynchpin that holds together the two metaphysical worlds of Cartesianism. As it had done for Aristotle, the *imagination/sensus communis* mediated between the bodily senses, and the (now incorporeal) rational mind.

3.1.3 Ideas

In the Empiricist tradition, images, under the rubric of *ideas*, came more than ever to be seen as the pre-eminent, even the only, form of mental content. Hume, indeed, frequently uses *imagination* as a virtual synonym for *mind*. This is reasonable as he regards the mind as no more than a *bundle of images* (*impressions* and *ideas*), and cognition as a matter of their vicissitudes and associative interactions. But even for Hume, imagination has a special perceptual role to play.

In a sense, for Hume, these latter functions of imagination are defects of the human mind. Philosophical analysis (he argues) shows that our natural belief (engendered by imagination) that the world contains various objects of various kinds cannot be rationally grounded.

Kant assigns similar functions to imagination (*Einbildungskraft*), but without such a sceptical twist. Imagination makes knowledge of the phenomenal world possible, by *synthesizing the incoherent sensory manifold into representational images suitable to be brought under concepts*. The Kantian imagination, in this regard, seems, broadly speaking, to fill much the same role as the Aristotelian *sensus communis*. However, it has by now become clear just how poorly we understand how any such faculty could possibly work. Kant rejects Hume's purely associative account, and argues that the imaginative synthesis must be governed by *a priori rules* or *schemata*, but how they operate is "an art concealed in the depths of the human soul, whose real modes of activity nature is hardly likely ever to allow us to discover" (Kant, 1998). The question as to how meaningful mental representations can arise in us remains obscure (or, at the least, very controversial) to this day.

3.1.4 Thought stuff

With the rise of the Romantic Movement in the latter part of the 18th century, the main focus of discussion concerning imagination shifted away from cognitive theory and epistemology, and towards its role in original, creative thinking, especially in the arts. This role had been acknowledged at least as far back as the work of Flavius Philostratus (c. 3rd century A.D.). Also, unsurprisingly, imagination had a long standing place in discussions of the nature of *dreams* and *visions*. What happened with Romanticism was not so much a re-conceptualization of imagination, but rather a revaluation of certain concepts that had long been associated with it. Originality, passion, the unreal, and non-rational thought, all generally deprecated in former ages, suddenly achieved a new, strongly positive evaluation.

Imagery played a large role in early experimental psychology (which, especially in Germany where it first flourished, was practiced as a branch of philosophy). Wilhelm Wundt, "the father of experimental psychology", founded the first psychological research and teaching laboratory in 1876, and imagery played essentially the same pivotal cognitive role in his theories (and those of most of his many students and imitators) that it had played for the philosophers of former ages.

Wundt's American counterpart, and contemporary, William James, took a not dissimilar view, although he was careful to acknowledge that in some people the "thought stuff", as he called it, might consist not so much of visual imagery but of imagery of other modes, especially the "verbal images" of inner speech. In his textbook *The Principles of Psychology* (1890 in James, 1967) James has much that is insightful to say about psychological processes in general, and about the role of imagery in them in particular, but, although he carried out experimental demonstrations in his psychology teaching at Harvard, James had little interest in the actual pursuit of experimental research, and established no graduate teaching program in experimental psychology. Anyway, he developed studies on associative thinking that can be regarded as forerunners of modern connectionism. In his chapter on 'Association' James considered someone thinking to a certain dinner party. He developed a scheme to represent an association network activated to remember the party, based on neural cell assemblies.

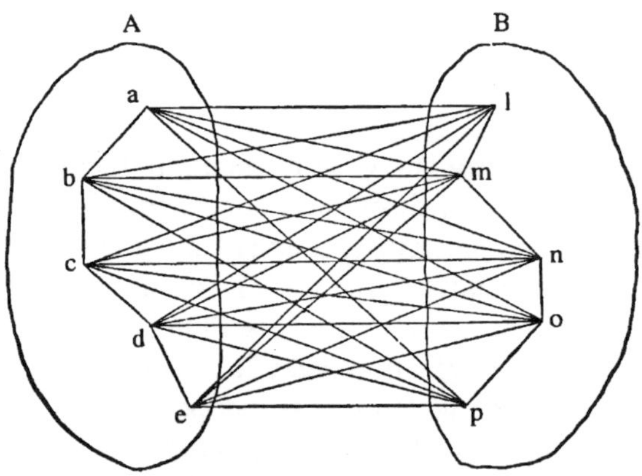

William James' scheme of a brain network (Mainzer, 1998)

3.2 Introspection *vs.* behavior?

Under the influence of both Machian positivism and, later, the act psychology of Brentano and the phenomenology of Husserl, Külpe, like Titchener (whom he had helped train), rejected what he saw as Wundt's unnecessarily strict methodological restrictions on the scope of empirical science. He encouraged his students to extend the scope of the introspective experimental method to the study of the higher processes of thought and reasoning. From about 1901 Oswald Külpe and his students at the University of Würzburg directly challenged these assumptions. Experimental subjects in Würzburg were asked to provide introspective reports of the contents of their consciousness as they performed specified cognitive tasks.

The introspectors (often Külpe himself, or other members of his research team) frequently claimed not to experience imagery, but rather *imageless thoughts*, conscious contents without any sensory or perceptual quality. In response, Wundt sharply criticized the introspective methodology of the Würzburg experiments, whilst Titchener, Wundt's leading disciple in America, reported that in his laboratory, similar introspective experiments always evoked imagery (not necessarily visual), producing no evidence whatsoever for conscious imageless thoughts. "Thinking in words" or the "inner dialogue" is plausibly regarded as a form of auditory or vocal-kinaesthetic imagery.

The bitter and irresolvable controversy that arose led to a reaction by which introspective methods became thoroughly discredited amongst the majority of experimental psychologists, and the notion of mental imagery fell out of intellectual favour. J.B. Watson, who inaugurated the very influential Behaviorist movement in psychology, questioned the scientific reality of consciousness in general, and imagery in particular.

The question of the reality of conscious imageless thoughts was left unresolved, and it remains so today. To borrow a coinage from Dennett (1978) "*Behaviorist psychology was thoroughly iconophobic*".

The analytical philosophy movement (that arose in the early 20th century, and that still deeply influences most English speaking philosophers) originated from the hope that philosophical problems could be definitively solved through the analysis of language, using the newly invented tools of formal logic. It thus treated language as the fundamental medium of thought, and several of the leading figures of the movement (notably Frege, Wittgenstein, and Schlick) argued strongly against the traditional view that linguistic meaning derives from images in the mind. These arguments were widely accepted, and imagery was relegated to the sidelines of philosophy. It no longer seemed to have a vital functional role to play in the workings of the mind.

Many psychologists and philosophers of this era came, partly for this reason, to feel that thought should be understood in terms of verbal language per se, and that it was a serious mistake ever to have thought that the representational power of language derives from that of some more fundamental form of representation, such as mental imagery.

Someone goes so far as to suggest (though without citing any evidence) that the work of the later Wittgenstein largely grew out of the reaction to the imageless thought affair. It may be overstating the case, but certainly a leading Würzburg alumnus, Karl Bühler, was established in Vienna during the inter-war years, and Wittgenstein is known to have met him there, and seems to have reacted strongly to his. Bühler also taught, and deeply influenced, the young Karl Popper, and undoubtedly his views would also have been quite familiar to the Vienna Circle positivists.

3.2.1 Revival

A revival of research on imagery was an important element of the cognitive revolution of the 1960s and 70s, contributing greatly to the rising scientific interest in mental representations. Seemingly, this revival initially stemmed largely from applied psychology research on *sensory deprivation* and on *hallucinogenic drugs*. Another important catalyst was Yates' seminal historical work on the significance of imagery *mnemonics* in ancient through Renaissance thought. Once the powerful mnemonic properties of imagery were experimentally confirmed, imagery could no longer be dismissed by psychologists. Interest was only heightened, during the 1970s, by the stunning *mental rotation* experiments of Shepard and his students, and experiments by Kosslyn demonstrating *mental scanning* and related effects. This work was taken to show that imagery is involved in *visuo-spatial reasoning*, and has inherently spatial properties.

But how could these findings on imagery be reconciled with the functionalist, computational symbol manipulation approach to cognition that was emerging during the same period? The standard philosophical interpretation of this approach depicts cognition as the computational manipulation of representations expressed in *mentalese* (the *Language of Thought*) a hypothetical, essentially language-like, representational system supposedly built into the brain (Fodor, 1975). Two rival approaches arose toward integrating the empirical findings about imagery into computational cognitive science.

Pylyshyn, in a series of influential papers (Pylyshyn, 1984), argued that all the genuine phenomena associated with imagery (indeed, all truly mental phenomena) can and must be explained entirely in terms of *mentalese* representations. For Pylyshyn and his allies, the computational paradigm of cognitive science demands that the underlying representational reality of imagery (and of actual perceptual experience) is *not picture-like*, but rather a detailed *mentalese description* of a scene.

Other cognitive scientists, however, notably Shepard and Kosslyn, argued that the evidence implies that imagery must be a distinct, *non-language-like form of representation*. Kosslyn, in particular, developed a *quasi-pictorial* computational theory of visual imagery, based on an *analogy with computer graphics* (Kosslyn, 1980). *Computer graphics files store information in a compressed, non-pictorial form, but when they are displayed they are translated into a mathematical map (bitmap) of the computer monitor screen, that specifies the color at each pixel (tiny dot) on the screen itself.* Likewise, suggests Kosslyn, visual information may be stored in the brain as compact descriptions, but we experience an image only when this information is used to create a two dimensional map of visual space in a special, functionally defined memory area he calls the *visual buffer*. The picture in Kosslyn's theory is merely quasi, because there is no equivalent to the monitor screen to display it. What we experience as imagery, and what is available to the cognitive processes that use imagery, is the functional picture, the mathematical map, in the visual buffer. In later work, Kosslyn (Kosslyn, 1994) identifies this *visual buffer* with the several retinotopically mapped visual areas of the brain.

3.2.2 Debates

Description and *quasi-pictorial* theorists disagreed sharply over what sorts of computational symbols, or data structures, are acceptable within cognitive theory, and which best capture the empirical properties of imagery. Nearly all philosophers accept that imagery has *intentionality*: the characteristically mental property of *being of, about or directed at some object* (real or imaginary). In this regard, mental imagery proper may be distinguished from superficially similar but non-intentional phenomena such as afterimages and phosphenes.

Most discussions of imagery, in the past and today, have indeed focused upon the visual mode. Nevertheless, there is every reason to believe that other modes of quasi-perceptual experience are just as common and important, and *imagery* has come to be the accepted scientific term for referring to them too: interesting studies of *auditory imagery*, *kinaesthetic imagery*, *haptic* (touch) *imagery*, and so forth, can be found in the contemporary psychological literature.

Indeed, there is some reason to think (although it is certainly not established) that that minority of people (about 10% of the population by some estimates) who deny ever experiencing imagery, or who deny that it plays any significant role in their mental lives, may simply be understanding the terminology in a somewhat idiosyncratic fashion.

Fundamentally, a *distributed representation* is one in which meaning is not captured by a single symbolic unit, but rather arises from the interaction of a set of units, normally in a network of some sort. In the case of the brain, the concept of 'grandmother' does not seem to be represented by a single 'grandmother cell,' but is rather distributed across a network of neurons (Churchland, 1995). This method of representation stands in direct opposition to the symbolic representation used by adherents of classical artificial intelligence (AI).

1. First, symbolic representations are strongly propositional. Thus, when this method of representation is used in non-language based applications such as image processing it becomes extremely difficult to explain many psychological findings. Similarly, taste, sound, touch, and smell are very awkward to handle with symbolic representations. In contrast, distributed representations are equally well suited to all modes of perception.
2. Second, symbolic representations are "all-or-none".
3. Furthermore, minor damage to a symbolic conceptual network causes loss of entire concepts, whereas a distributed network loses accuracy, as people do, not entire concepts.

4. The prototype for a concept would not be defined by one particular example, but rather a superposition of all available examples of the concept.

Distributed representations do not fall prey to these short-comings. Rather, distributed representations:

1. Are the natural result of organization of statistical input and thus provide a natural means to capturing semantic information;
2. Have been successfully applied to visual, olfactory, auditory and tactile problems;
3. Have been proved to degrade gracefully with noise and are commonly tested with simulated lesions (i.e. a removal of part of the representation).

Furthermore, distributed representations have a number of properties which lend significant psychological and neurological plausibility to models employing them. In many instances distributed representations:

1. Represent concepts continuously;
2. Can be processed in parallel;
3. Can be learned using proven methods.

These powerful properties, coupled with the ability of distributed representations to overcome important short-comings of symbolic representation provide a solid foundation for realistic computational models of human cognition.

Though many dynamicists feel that connectionism should be replaced by dynamicism, it is not clear why. Connectionist networks exhibit many of the behaviors that dynamicists note as being central to cognition and incompatible with classical, or symbolicist AI.

3.3 Re-Presentation

John Locke used the term to indicate that *words stand for, or re-present ideas*; occasionally he used it also for the mental image of an idea, either fictitious (e.g. "the unspeakable joys of heaven") or "made up of ever so many particulars". It was presumably this second meaning that led translators of Kant's Critique of Pure Reason to use *representation* for the German word *Vorstellung*. This was unfortunate because the ordinary-language meaning had practically superseded the Locke's more abstract second meaning.

The German word *Vorstellung* indicates a range of things that take place spontaneously in someone's head and do not require an original. In contrast, the English *representation* normally refers to sensory material that more or less reliably evokes a common experience. That is to say, it is used to call forth in the beholder specific remembered situations, events, or objects (as for instance, the flowers and fruits in Matisse's drawings). The term, therefore, indicates a perceptual cause whose effect is expected to be an image in the mind of the perceiver. In short, *representation* in English inevitably suggests reference to an original, in much the same way as does copy, replica, and reproduction. Note that *représentation* in French and *rappresentazione* in Italian function in the same way.

3.3.1 Things

In contemporary psychology, and in the study of intelligence, be it natural or artificial, the term representation is used for mental images that are supposed to reflect, or correspond to things that lie beyond our experiential interface. Jerry Fodor, for instance, explained in his interview with Baumgartner and Payr (Baumgartner, 1995): "mental states that represent states of the world" constitute a representational capacity.

In another interview in the same collection, James McClelland says: "I agree with Newell: the mind is a physical device that performs operations on physical objects, which are not the things themselves but representations of those things". Other English speakers in the field of cognitive science have also explicitly or implicitly subscribed to the view that representations represent real things. It does not matter whether they define "to represent" as to depict or to stand for, because in both cases there is no way of checking any such relation to "real" things.

The same goes for neurophysiologists who speak of observable phenomena in the brain as "representing" certain activities and results of the mind.

This use of representation is misguided, because it entails the belief that certain ideas we abstract from our experience correspond to a reality that lies beyond experience. It would be interesting to study the reasons why this belief manages to survive in the face of a long history of incontrovertible refutation. It has survived in spite of the fact that it was shown to be illusory by Xenophanes at the time of the Pre-Socratics; and it survived in spite of all the sceptics since, who have amply demonstrated the logical impasse that makes it illicit to claim a representational relation between mental images or structures and any independent external reality.

3.4 Presentation

The main argument of the sceptics is simple and irrefutable. To know whether anything we derive from experience corresponds to, or "represents" an aspect of an external world, we should have to be able to compare it to the real thing. But this we cannot do, because we can compare experiences only to other experiences. It would be wiser to scrap the term *representation* in these contexts and to follow Mark Baldwin's example and speak of *presentations*. This would be much closer to the Kantian term *Vorstellung*. These *presentations* are pieces of experience that we have combined in order to form more and less complex structures, in our attempt to order and systematize the world in which we find ourselves living. It is the only world we know and it's a world that only we ourselves perceive and conceive.

Insofar as we remember these structures, we can recall them and then they are *Re-Presentations*. The most obvious instances of re-presentations happen in our *dreams*, when there is no perceptual activity at all.

You may, for example, dream that you are in a room, but all you see of the room is a door (perhaps because you expect someone to come in through it). You have no idea of the size of the room, and there are no windows, curtains, pictures, no ceiling or furniture, or anything else that usually characterizes a room. These items may come in later -- as the plot of the dream develops -- but at this point, they are irrelevant in your dream-presentation of a room. In contrast, your perception of a room starts from sensory impressions that you proceed to coordinate, and they then allow you to consider them compatible with your interaction with the "room". This interaction co-evolves as a presentation.

One can therefore say: in perception, sensory signals call up a concept, in re-presentation, on the other hand, a concept calls up sensory impressions. In neither case is the experience caused by what philosophers want to call *reality*.

3.4.1 Imago

Within semiotics, the term icon in rarely used in its most common religious and art historical sense, to refer to a pictorial representation of persons or events derived from the sacred history of Christianity, particularly as used as an aid to devotion, although the only extant semiotic monograph concerned with a single pictorial genre is in fact about icons in this sense.

Nor is the term normally used to refer to all things visible or everything they elements of which are graphically disposed, as in the jargon of computer programming, or in cognitive psychology. In semiotic jargon, derived from Peirce, an icon is a sign in which the *thing* serving as expression is, in one or other respect, similar to, or shares properties with, another *thing*, which serves as its content. In fact, if we follow Peirce, there are two further requirements: not only should the relation connecting the two *things* exist independently of the sign relation, just as is the case with the *index*, but, in addition, the properties of the two *things* should inhere in them independently.

Thus, icons in the religious sense are not particularly good instances of icons in the semiotic sense, for they are, as Uspenskij has shown, subject to several conventions determining the kind of perspective which may be employed, and the kind of things and persons which may be represented in different parts of the picture. Contrary to the icons of computer programs and those of cognitive psychology, *iconic signs may occur in any sense modality*, e.g. in audition, notably in verbal language (not only onomatopoetic words, but also in the form of regularities and symmetries) and music. Not all visual signs are iconic in the semiotic sense: indeed many icons found in computer programs are actually aniconic visual signs.

Many semioticians, in particular those who deny the existence of iconic signs, apparently believe pictures to be typical instances of this category. There are several reasons to think that this was not Peirce's view. Pure icons, he states, only appear in thinking, if ever. According to Peirce's conception, a painting is in fact largely conventional, or "symbolic". Indeed, it is only for a floating instant, "when we lose the consciousness that it is not the thing, the distinction of the real and the copy", that a painting may appear to be a pure icon (Sonesson, 1989).

3.4.2 Sensible objects

It will be noted then that a pure icon is thus not a sign, as the latter term is commonly understood. At first, it may seem that, although the icon is not a socially instituted sign, i.e. not something which is accepted by a community of sign uses, it could at least, for a short time span, become a sign to a single observer. But even this is contrary to the very conditions described by Peirce: he specifically refers to the case in which the sign loses its sign character, when it is not seen as a sign but is confused with reality itself (which could actually happen when looking at a picture through an key-hole with a single eye), when, as Piaget would have said, there is *no differentiation between expression and content*.

Indeed, if would seem that, at least sometimes, the pure icon is taken to be something even less substantial: an impression of reality, which does not necessarily correspond to anything in the real world, for "it affords no assurance that there is any such thing in nature". Thus, it seems to be very close to the *phaneron*, the unit of Peircean phenomenology (itself close to the Husserlean *noema*), which is anything appearing to the mind, irrespective of its reality status. In this sense, the Peircean icon is somewhat similar to that of cognitive psychology, for it involves *sensible objects*, not signs in any precise sense: however, it still comprises all sense modalities.

In most cases, when reference is made to icons in semiotics, what is actually meant is what Peirce termed *hypo-icons*, that is, signs which involve iconicity but also, to a great extent, indexical and/or symbolic (that is, conventional, or perhaps more generally, rule-like) properties. There are supposed to be three kinds of hypo-icons:

- *images*, in which case the similarity between expression and content is one of "simple qualities";
- *diagrams*, where the similarity is one of "analogous relations in their parts";
- *metaphors*, in which the relations of similarity are brought to an even further degree of mediation.

Diagrams in the sense of ordinary language are also diagrams in the Peircean sense, e.g. the population curve which rises to the extent that the population does so (Peirce, 1931).

3.4.3 Peircean iconicity

Contrary to the way in which icons have been conceived in the later semiotic tradition, diagrams, rather than pictures, are at the core of Peircean iconicity: at least, they are of most interest to Peirce himself. Indeed, mathematical formulae and deductive schemes, which are based on conventional signs, are those most often discussed in his work.

There is still another sense in which pictures are far from being central instances of icons. As was noted above, the fact that an object serving as the expression of an icon, and another object serving as its content, possess, in some respects, the same properties, should not be a result of one of them having an influence on the other. In the case of an icon (contrary to the case of an index), "it simply happens that its qualities resemble those of that object, and excite analogous sensations in the mind for which it is a *likeness*". Since both Franklin and Rumford are Americans, Peirce claims, one of them may serve as a sign of the other; but the fact that Franklin is an American is quite unrelated to Rumford's being one. But there is at least one sense in which this is not true, not only of a photograph (which Peirce often pronounces to be an index), but also in the case of a painting or the image on a computer screen: in each case, the "thing" serving as the expression is expressly constructed in order to resemble the "thing" serving as the content, although a direct physical connection only exists in the first instance. Leonardo painted the canvas known as Mona Lisa in order to create a resemblance to the wife of Francesco del Giocondo, and, although the resemblance is of a much more abstract kind, the same is true of Picasso painting Gertrude Stein. And it is as true of a synthetic computer picture showing a lamp as of a photograph with the same subject.

A car, which is not a sign on the street, becomes one at a car exhibition, as does Man Ray's iron in the museum. We have to know the show-case convention to understand that the tin can in the shop-window stands for many other objects of the same category; we need to be familiar with the art exhibition convention to realise that each object merely signifies itself; and we are able to understand that the tailor's swatch is a sign of its pattern and colour, but not of its shape, only if we have learnt the convention associated with the swatch (Sonesson, 1989).

Convention is thus needed, not only to establish the sign character, but also the very iconicity of these icons. Since iconicity can be perceived only once the sign function, and a particular variety of it, is known to obtain, the resulting icons may be termed secondary.

Such a primary icon is actually realised by the picture sign. Indeed, we know from child psychology and anthropology that no particular training is needed for a human being to perceive a surface as a picture (cf. picture perception). The possibility of this feat remains a mystery: they properties possessed in common by the picture and that which it represents are extremely abstract. It has been suggested that picture perception is only possible because there is a taken-for-granted emergence of properties in the world of everyday life which makes certain objects and materials more probable sign-vehicles than others (Sonesson, 1989).

3.5 Index

Together with the icon and the symbol, the index constitutes the three kinds of signs distinguished by Peirce according to the relation contracted by the expression (Peirce's *representamen*) and the content and/or referent (Peirce's *object*, i.e. *immediate* or *dynamic object*). According to one possible paraphrase, which seems sufficiently broad to account for most of Peirce's examples and a fair amount of his definitions, *an index is a sign in which the 'thing' which serves as the expression is*, in one or other way, which is independent of the sign relation, *connected with another 'thing', which serves as its content* (indexicality). The term, if not the concept, has, however, a number of other sources, some of which are not necessarily compatible with Peirce's conception, and not all uses of the term, neither by Peirce, nor by later semioticians, may really concern the same concept.

The term chosen by Peirce certainly suggests that *all indices, like the pointing index finger, or an arrow, serve to pinpoint a particular object, to isolate it and bring it out of the, typically spatial, context into which it is ordinarily enmeshed*; and this is indeed what Peirce affirms. However, if we introduce the term *indicator* to describe signs which are employed to single out an object or a portion of space for our particular attention, it may be argued that they are not necessarily indices in Peirce's sense, and that they are not, in any event, sufficiently characterised by being so classified (Sonesson, 1989). Thus, certain indicators, such as pointing fingers and arrows, do suppose a *relation of contiguity* with that which they point to; but this is not necessary, or even possible, in the case of many verbal indicators, most maps, and the photographic options depending on film, lighting, and frame described as indexical in the semiotics of photography; for, in these cases, the indicative gesture is merely recreated at the level of content. At least some of these examples would also be described by Peirce as not being *genuine indices*.

On the other hand, real indicators, such as fingers and arrows, are equally contiguous to a number of objects which they do not indicate, for instance to the things which are at the opposite side of the arrow-head, in the direction to which it does not point. Therefore, *something beyond mere indexicality is required, in the case of the arrow, for instance, the forward thrust of the arrow-head as imagined in water, or the sentiment of its slipping from our hands*, as Thom (1975) has suggested. To term certain signs 'indicators' is, obviously, to make a categorisation of signs on the basis of their functions, as seen in relationship to the over-all scenes in which signs are produced. We should not expect this categorisation to coincide with the one stemming from Peirce's classification, which depends on the nature of the relationship between the expression and the referent or content of the sign. Of course, from this point of view, the term 'index' is a misnomer, for although the finger so termed may function as an index, its specific function goes beyond that.

3.5.1 Indexes

Another sense of index current in semiotic literature, which has no obvious relation to that of Peirce, is the one employed by Prieto. In order to describe how signs come into being, he imagines an elementary situation in which a set of *hoof-prints* made by a horse may be observed. Something which Prieto calls the '*significative indication*' tells us that there is a horse around. There is not, in this case, as there would be in a linguistic sign, any *notificative indication*, which would convey to us the idea corresponding to the phrase "attention! this is intended to contain a message". In this particular case, the hoof-prints would be indices also in Peirce's sense, but not for the same reason: not because they are non-intentional, but because of being markings on a surface. In the case of a linguistic sign, however, the *significative indication*, which corresponds to the perception of the sound, would not ordinarily be indexical.

There is another usage of the term 'index' which Piaget employs when introducing his idea of the *semiotic function* (which, in the early writings, was less adequately termed the symbolic function), which is a capacity acquired by the child at around 18 to 24 months of age, enabling him to *imitate* something outside the direct presence of the model, to use language, make drawings, play "symbolically", and have access to mental imagery and memory. The common factor underlying all these phenomena, according to Piaget, is the ability to represent reality by means of a signifier which is distinct, or differentiated, from the signified. Even before that age, however, Piaget believes that the child is able to "connect significations" by means of "indices" and "signals", which do not suppose any such differentiation between expression and content. The signifier of the index is, Piaget says, "an objective aspect of the signified"; thus, for instance, the visible part of an almost entirely hidden object is the signifier of the object for the baby; and the tracks in the snow stand for the prey to the hunter, just as any effect stands for its cause. But when the child uses a pebble to signify candy, he is well aware of the difference between the two, which implies, as Piaget tells us, "a differentiation, from the subject's own point of view, between the signifier and the signified".

It is important to note that, while the signifier of the index is said to be an objective aspect of the signifier, we are told that in the sign, expression and content are differentiated form the point of view of the subject. We could actually imagine this same child that in Piaget's example uses a pebble to stand for a piece of candy having recourse instead to a feather in order to represent a bird, without therefore confusing the feather and the bird: then the child would be using the feature, which is objectively a part of the bird, while differentiating the former from the latter from his point of view. Only then would he be using an index, not in the sense of Piaget, but in that of Peirce and most semioticians, rather than a mere indexicality; and obviously the hunter, who uses the tracks to identify the animal, and to find out which direction is has followed, and who does this in order to catch the animal, does not, in his construal of the sign, confuse the tracks with the animal itself, in which case he would be satisfied with the former (Sonesson, 1989). As the term is employed by Piaget, the index is thus not a sign: it may therefore better be termed indexicality, or an indexical ground, which as yet has not attained the status of sign. When suitably defined, however, Piaget's notion of differentiation may be used, together with other criteria, to distinguish signs from other meanings, and thus to tell indices and other indexicalities apart.

Apart from being a sign, an index, in the Peircean sense, must contain an indexical ground (cf. indexicality). The fact that such a ground could exist independently of the sign relation should not be taken to mean that the indexical relation necessarily has to precede the sign relation in real time. Indeed, some indexical relations must come into being at the same time as the sign is produced, as is the case, for instance, of *verbal shifters*: the person indexically related to the sign 'I' is the one which at a particular moment pronounces the sound /ai/, which is to say that the indexical ground is produced at the very same moment in which the sign is put to use. There are no class of "pointed-out objects" known to exists, but a member of such a class is created each time an act of pointing takes place.

3.5.2 Abduction

Many of Peirce's own examples of indexical signs are of the kinds which acquire their meaning thanks to a regularity which is known to obtain between different facts. Since a kind of reasoning which connects two facts by means of supposed regularity is called an *abduction* by Peirce, we might perhaps call this group of signs *abductive indices*. They can involve contiguity, as in the case of footprints, fingerprints, the cross as a sign of the crucified, the weather-cock (contiguity to the direction of the wind); or factorality, when an anchor is used to stand for navigation, the clock to designate the watch-maker's (as part of the sum total of clocks), or a painting to indicate the painter's workshop.

Some of Peirce's examples, and many of those suggested later, are however of another kind for, instead of presupposing a regularity known to obtain between the "thing" which serves as expression of the sign, and another "thing" which is taken to be its content, they transform something which is contiguous, or in a relation of factorality, to the expression, into its content. These signs may therefore be termed *performative indices*. With contiguity, they give rise to such phenomena as the pronoun 'you', the finger pointing to an object, the weathercock (as marking the here-and-now of the wind), the clock of the watch-maker's (as marking the emplacement of the shop); and with factorality, they may produce the pronouns 'I', 'here', 'now', the finger pointing out a direction, etc.

3.5.3 Semiotic streams

Finally, many *secondary signs*, or signs standing for other signs, are often said to be indexical. Secondary indexical signs are signs, the dominant sign relation of which is pictorial or otherwise iconical, conventional, or whatever, but which involve a secondary sign function, which is a relation between several signs, or parts of several signs, which is indexical. The most obvious case of such signs are the rhetorical figures of *metonymy* and *synecdoche*, in which the entire primary sign is related to another sign by means of the respective contents. More trivial are such examples in which the primary sign content is related to a secondary content outside the sign (implication, connotation in Eco's sense, metonymy and synecdoche in a loose sense, etc.), some examples of which are dead rhetorical figures such as the sword for the army, a picture of a cross for the crucified, the traditional "symbols" of iconography (all with contiguity); dead rhetorical figures such as the sail for the ship, a picture of a clock for the watch-maker's, iconography generally (all with factorality).

Another variety, which is often parasitic on *pictorial signs*, is when the entire primary sign, which are related to another sign via their respective expressions, form an actual perceptual context. This type is often found in publicity and in surrealist painting. Some examples, with contiguity, would be two figures seen against a common ground, for instance an advertisement for a brand of whisky with a glass and a bottle; a bottle of gin and a crown; a jetty and a tyre; or pictures placed side by side as in collages, montages, etc. With factorality, examples would be a pile of fruits forming a crown; Magritte's drawing of a face which is also a female trunk; slices of orange forming a jam bottle; Arcimboldo's portraits, etc. Other varieties may also be distinguished (cf. Sonesson, 1989).

Peirce introduces a distinction between different kinds of indexical signs, but in terms which are not easy to interpret: thus, there are designations, which stand for "things or quasi-things with which the interpreting mind is already acquainted", and reagents, which serve "to ascertain certain facts", based on our knowledge of the "connection with the phenomena it indicates". Examples of the first kind are personal, demonstrative and relative pronouns, as well as proper names; an instance of the second kind, however, is when "water placed in a vessel with a shaving of camphor thrown upon it will show whether the vessel is clean or not."

3.6 Psychoanalysis

Stuart Kauffman writes about psychoanalysis that in the beginning of this century it burst upon the intellectual scene, but at the end of the century it is held in disregard: "unscientific, untestable, ineffective and expensive", just the opposite of pharmacology. But the truth could be deeper and more complex, as he asserts that in a decade "psychoanalysis will be regarded as the forefront arena of science" (Palombo, 1999). He cites many complex phenomena of bio-psycho-social interaction in which psychoanalysis can bring seminal contributions to the new frontiers of life sciences. Many of these contributions are related to one of the main missions that psychoanalysis has assigned to itself since the beginnings: the study of iconic thinking.

Laplanche and Pontalis (1968) following the place of imagery in the psychoanalytical works attempt to define its structural aspects. They conclude that the *phantasma* is intrinsically connected with desire and prohibition, as it is meant by its German signifier *Wunschphantasie* and its Jewish antecedent *yetser*. In the beginning the *phantasma* in psychoanalysis is anthropologically based on some *Urphantasien*, primal scenes, in which the theatre of desire is a drama (comedy or tragedy) about the origins of the desiring subject. In these scenarios the imagination (*Einsbildungskraft*) becomes iconic creation (*Das Phantasieren*).

3.6.1 Wunschphantasie

Wunschphantasie brings within itself the *intentionality* that Brentano, influencing the young Freud, who was between his students, posed in the foundations of psychoanalysis. We find here also the impingements of other classical philosophers: the *little perceptions* without awareness of Leibniz; the *obscure representations* in Kant; the *sunken ideas* of Herbart; and the *unconscious conclusions* of Helmoltz. *All these interstitial concepts are the forerunners of the Id (It), the other subjects we are lived by, as the vector of consciousness sometimes called the Unconscious.*

The young Freud in his "G-Draft" to Fliess schematized in a topological way some mind processes as in a vector-field (Freud, 1954): a phase portrait. This scheme is connecting, in a circuit of desire, the subject to the other, on the landscapes of the external world, the inner world, and the biological body.

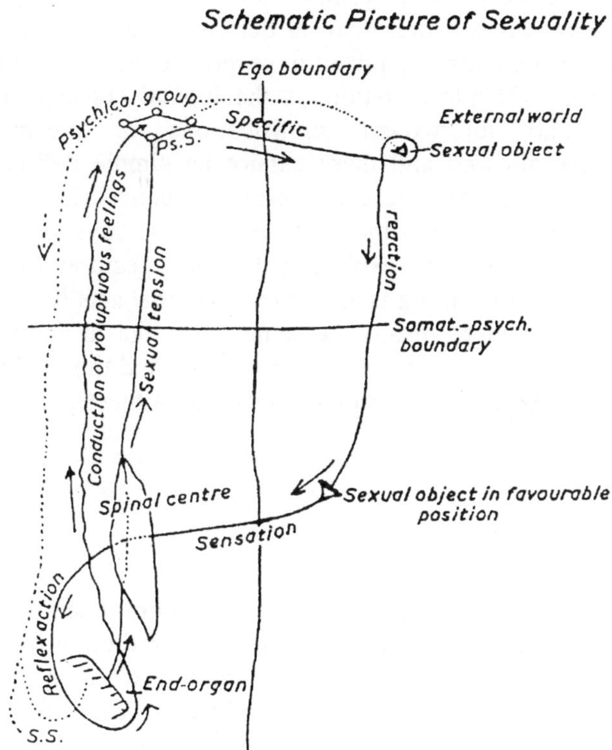

G-Draft scheme (Freud, 1985)

The elements of the model are allocated in a space divided by the intersection of two boundaries, the Ego boundary, and the psychosomatic border: 1) an object, in the external world; 2) an object in a favorable position, outside the Ego, in the body; 3) an end-organ, a somatic source, and a spinal center, in the body-Ego; and 4) a psychic sexual group, in the psychic Ego. These elements are the main stations of a circuit performed by a vector: sexuality, with its drive, goal, source and eventual obstacles. It is amazing how this dynamical and complex model has been neglected in the history of psychoanalysis and psychosomatics. This model is entirely based on drive, movement and loops, while other famous Freudian models are more deterministic and static.

Freud was well aware of the complexity of mind dynamics and he coined the specific term *Uberdeterminierung*, or over-determination: every unconscious formation is a multi-causal derivative; and every derivative is linked to multiple elements, organized in several sequences of meaning. The *Uberdeutung*, the over-interpretation, is the logical consequence: in these non-Euclidean and interwoven stratifications of meaning, several interpretations are allowed and there cannot be simple defined criteria to define the correct one. Are these properties of connections in some way related to the semiotic nature of the knots in the associative network?

In same years Peirce was exploring the same features of the semiotic universe asserting that meaning is an *interpretant* that acts like an interpreter who asserts that a stranger means the same thing, but in another language. Every interpretant may have its own interpretants, and this yields an infinite chain of possible translations (symbol-interpretant-object): an endless fugue in the semiotic universe.

3.6.2 Other subjects

Regarding the status of unconscious representations Kant had raised the question, against Locke: there is an apparent contradiction on unconscious representations: how can we think about them if we aren't conscious of them? Freud was aware of this theoretical problem because there is a part of the (original or primary) unconscious, or the *Es*, as he will call it later that is similar to the Kantian *Ding an Sich*: it is unknowable in itself, and we can just know some preconscious derivates, presentations. It is like in dreaming: we aren't able to know the original dream, *the dream-in-itself*. We can just know the conscious narration of a supposed dream. This will give Wittgenstein the start for many interesting notations in his reflections on the philosophy of psychology. This first *Unconscious*, or later *Es*, is someway rooted in social and biological codes. Freud in his Metapsychology (1924) was defining two different kinds of representations: word-representation (*Wortvorstellung*) as a mnemonic trace of verbal language and thing-representation (*Dingvorstellung*) as iconic trace of perceptions or imagination.

In the Studies on Hysteria there is a suggestion that the cathartic method of the *talking cure* could depurate mind from imagination by verbalization: "I was freeing her from all the phantasms she accumulated since my previous visit". The *phantasma* could be seen as traumatic by itself? Here the Freudian theory could bear the prejudice against phantasy from the Jewish tradition or from Vienna Circle, and its privilege for verbal thinking.

The scenarios of imagination in which the subject is representing its origins are defined as *Urszenen,* the primal scenes of origins. A basic feature of these scenes: it is almost impossible to establish if they are based on previous perceptions (of real events, eventually modified), or they are entirely built by the subject's mind forgery.

About this ambiguity Ernest Jung suggested that phantasies could be built during adult life and dated backwards: *Zuruckphantasieren* is the word he used for this activity. Freud agreed that later experiences may restructure memories in a backwards way He called this *Nachtraglich* effect, the *apres-coup* in Jacques Lacan. Freud was sceptic about Jung's solution and wrote that the regularity of these scenarios can suggest some sort of genetic or socio-anthropological transmission of this symbolic capacity.

3.7 Urszenen

The Urphantasien represent a connection between body, mind and culture, if in this term we include social structures, laws and genealogies. The personifications, *dramatis personae*, form a personal legend in which the subject can represent itself even in different positions at the same time. *In these scenes the Es is "the impersonal other subjects inside ourselves, the aliens who make us know by means of intention and desire".*

This structural function of the iconic language is explored in Levi-Strauss, who examined hundreds of indigenous myths through oppositional analysis in order to reveal their underlying systems (Lévi-Strauss, 1964).

Mythologies structure and give orders to our thinking: *"myths think in men without their knowing."*

Levi-Strauss criticized Jung's idea that a given mythological pattern, the so-called archetype, may have a certain meaning. It would be comparable to the long-supported error that a certain sound may possess a certain affinity with a meaning: for example the *liquid* semi-vowels with water etc. Now the *arbitrary character of linguistic signs* is considered as a prerequisite for a scientific study of language.

The meaning is in structures as clusters of relations. We should be able to read the harmony in the whole to find the meaning of these clusters, synchronically and diachronically.

Levi-Strauss applied this methodology to hundred of mythological cycles. We would exemplify his methodology by recalling his analysis of the Oedipus cycle, which has been considered so important by traditional psychoanalysis. But it has been oversimplified and highlighted just in some components, maybe not the most important. The genealogy of Oedipus, his blood relations with Dionysus and the Asian world, the meeting with the Sphinx: these are just a few parts of the mythological cycle left in shadow by Freud.

BLOOD RELATIONS -	***BLOOD RELATIONS +***	***AUTOCHTHONOUS ORIGIN -***	***AUTOCHTHONOUS ORIGIN +***
Cadmos seeks his sister Europa, ravished by Zeus			
		Cadmos kills the dragon	
	The Spartoi kill one another		
			Labdacos (Laios' father) = lame (?)
	Oedipus kills his father, Laios		Laios (Oedipus' father) = left-sided (?)
		Oedipus kills the Sphinx	
			Oedipus = swollen-foot (?)
Oedipus marries his mother, Jocasta			
	Eteocles kills his brother, Polynices		
Antigone buries her brother, Polynices, despite prohibition			

Mythological clusters (Lévi-Strauss, 1958)

If we consider this table representing a larger part of the Oedipus Cycle we find a diachronic order but also another subtler order referring to origins: *blood relations* and *autochtonous origin of mankind.* (The idea that men can be born from earth is present in various cultures: from Adam in the Bible to the Zunis in the Americas.) While in the first column there is an overrating of blood relations, in the second one there is an underrating. In the third column we find monsters coming from earth that have to be killed to allow the mankind's birth. In the last column we find Oedipus and his ancestors and we discover on the roots of their names' that they all have problems in walking: detaching and separating from earth. It might be recalled that he is also cousin of the tellurian god, *Dionysus*, who is called also *loxias*, "the obliquely walking": *the god of merging and emergence.*

All this mythological cycle is organized in a matrix of possible answers to the problem of the emergence of human subjectivity. The question from the Sphinx to Oedipus was just about this.

After the analysis of many mythologies (including the Zuni myth of emergence) Levi-Strauss suggests that a temporary conclusion regarding a general structure of myths can be the following:

$$F_x(a) : F_y(b) \cong F_x(b) : F_{a-1}(y)$$

It is assumed that a relation of equivalence exists between two situations defined respectively by an inversion of terms and relations, under two conditions: that one term be replaced by its opposite; and an inversion is made between the function value and the term value of two elements.

The myth needs two events to be construed: the first is just happening as an index (contiguity); the second is re-presented as an icon. This is the core of mythologies, as individual consciousness and social culture are.

3.7.1 Polymorph

The personal mythology *par excellence* is the definition of the Ego. As Glover noted (1970) "the earliest ego tendencies are derived from numerous scattered instincts": *the ego of a five year old is not only multinuclear, each nucleus is itself a polymorph.* The raw material of a possible early ego formation is on one hand various body areas of libidinal and aggressive powers: skin, oral, respiratory, excretory, muscular, and visual. The cluster formation of nuclei represented by aggregates of imagery activities and sensual-motor schemes organized around strategic areas of the body-mind matrix. It is a fractal disordered organization (or ordered disorganization); a multitude of fragmentary sources of intentionality. Each of them brings an *episteme*: an activity for possess and knowledge. It is an activity that evokes and meets objects.

The contribution of W. R. Bion is mainly related to the *transitions* between different semiotic states. He suggested that there can be transformations between indexes, icons and symbols and that these transformations follow rules and are context-dependent. He used an idiosyncratic terminology to define these semiotic entities: called the raw sensory data (that can be equated with indexes) β *elements*, while the precursors of icons were called α *elements*. He ascribed a possible transformation from β to α as due to a psychological function he called *reverie* that can be rooted in the *imaginary* functioning of the mind. There is an interesting schematization he proposed about these transitions (Bion, 1992).

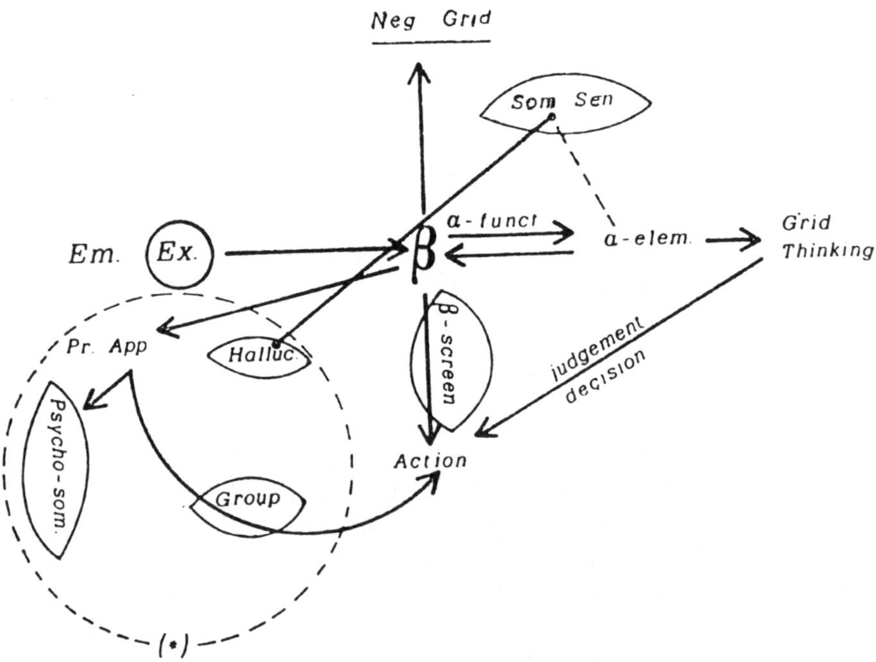

In this scheme Bion suggests that in some specific contexts, like group interactions, psychotic hallucinations, operative actions, emotions and psycho-somatic disorders, there can be a discharge of indexes. There is a transition between raw sensory data and iconic imagery. In other contexts the reverie could reverse the prevalence. Bion sees iconism as a precursor of symbols and proposes a classification of sign that he calls the Grid.

There can be a hierarchical privilege in this view, in which symbols are seen as "less primitive" than indexes (β elements in Bion's terminology). This could lead to the absurd consequence that the ideal condition could be seen as a complete transformation of all the β elements at least in α, if not all the semiotic entities in symbols. It can be considered a colonialist approach to semiotic life (mind), which has premises in Freud when he wrote: "*Wo Es war soll Ich werden*" ("where there was the Id there will be the Ego"). Freud was inspired by the contemporary dry of the Zuydersee marsh in Austria, whose land was conquered from "wild Nature" and destined to human agriculture.

Bion was aware of the limits he had to deal with when he complained that there wasn't available mathematics for the problems he studied (mind), and was trying to find it reading Poincarè's works.

He also suggested, regarding the possibility to have the pure experience of raw data, that in some conditions it might be useful a reversal of the *reverie* function, because the *ding-an-Sich*, can be found only on the other side, after the indexes of the β elements, nearby the hallucination.

The risk of a colonization of the "wild forests and lakes of the mind" is always at hand, because they host resources and dangers. Why not respect a sustainable and optimal balance between the complex different components of semiotic life? How can we assert that indexes are "more primitive" than icons and symbols? Information theory is the proof of the function that indexes have for knowledge.

3.8 Transitional phenomena

The works of Donald Winnicott opened new routes. From his discovery of the so called *transitional phenomena* and *objects*, the precursors of signs in developmental psychology, he highlighted these complex interactions. *Transitional phenomena* are ways to master the absence of the reference object of love, meaning and attachment. They represent a crucial emotional and cognitive transition to access the universe of language (Winnicott, 1971).

Child	***Thumb***	***Transitional Object***	***Type of Child***
X Boy	0	Mother, Rabbit (comforter)	Mother-fixated
Y Boy	+	'Baa', Jersey (soother)	Free
Twin A, Girl	0	Dummy, Donkey	Late maturity
Twin B, Boy	0	'Ee', Ee	Latent psychopath
Child of Y, Girl	0	'Baa', Blanket (reassurance)	Developing well
Child of Y, Girl	+	Thumb, Thumb (satisfaction)	Developing well
Child of Y, Boy	+	'Mimis'*, Objects (sorting)	Developing well

* Innumerable similar soft objects distinguished by color, length, width, and early subjected to sorting and classification (partially modified from Winnicott, 1958).

This is the original table that Winnicott used to schematize the transitional phenomena he observed at the junction of attachment and language development.

He also used a peculiar method for therapeutic consultations in child psychiatry based of the interactive production of squiggles:

"Now I made a squiggle and he made it into 'a snake curled up'. 'It is happy'. He gave this drawing a certain amount of attention and then he said quite spontaneously: 'I like the snake that is curled up'. While this was going on he was beginning to finger his face and to play with his face with the pencil. I noted the feeling of a relationship between the ideas of the curled up snake these infantile relics showing in his handling of his face. I also remembered that his mother had told me that when very young instead of using a transitional object he had tended to need her actual face the skin of which he would stroke until he went to sleep. (...) I did refer to the happy snake in terms of himself curled up on his mother's lap, feeling safe and protected from the world (...) a statement made in terms of going out into the world and growing up versus regressive dependence." (Winnicott, 1971b).

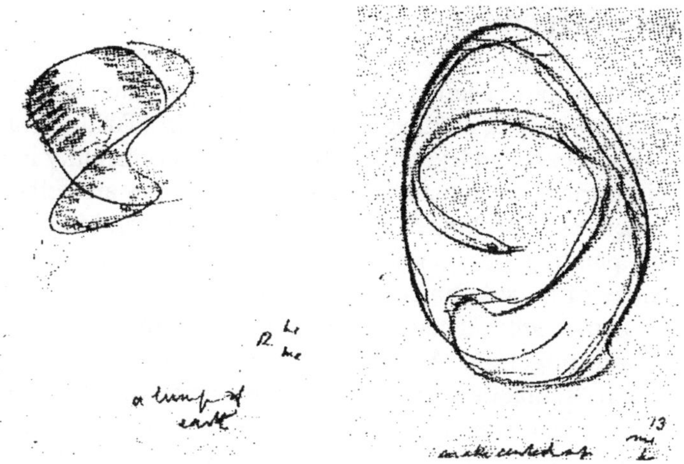

The "curled snake" squiggles (Winnicott, 1971b)

The transcription of the sequence of verbal and motor interaction (as represented in the squiggle) is very rich of implications for our exploration. *The squiggle technique brings embedded in itself motor activity, perception and the genesis of icons with added verbalization. It is generative of a rich and complex small semiotic universe. You can appreciate its freshness in the instants of creation.* So rich and fresh that sometimes even Winnicott's interpretations seem added from another, external, world (Stein, 1998). As if in the presumption of master and control instead of synchronize and co-evolve. The squiggle game is a laboratory for the microevolution of language. In the example we have cited we find indexes (gesture), tactile icons (perceptual shapes designed by movement on the skin), visual icons, and symbols (words as descriptions or metaphors). We can notice the transitions between these different semiotic areas.

3.9 Autistic shapes

The perceptual shapes recalled from this "squiggle interaction" remind the *autistic sensation shapes* studied by Frances Tustin (Tustin, Frances, 1990) that can be found in the indexical functioning of every healthy individual and become so important for autistic subjects. These shapes are self-generated by behaviors such as "stroking, brushing, smearing, rubbing and even by drawing and painting on smooth surface, both those of the subject's own body and those of outside objects experienced as part of the subject's body surfaces" (Tustin, 1990). They are also derived from so-called mannerisms as rocking and fidgeting. As autistics tend to live on two dimensional experiences, on "flatlands", they avoid angles and everything could remind to depth. They find soothing and tranquillizing the construction of flat shapes and icons, flat worlds. Thus, painful awareness of bodily boundary and separateness is avoided by finding shapes without the third dimension of a possible distance (Freeman, W. J., 1986). Autistic children talk about "their shapes" and if they are psychologically testable they are very good with the matching of shapes. If the autistic shapes can be completely idiosyncratic they rely on the shaping function of human mind, on the primal sensory-motor schemes permanent source of all icons and *Gestalten*.

3.10 Embodied cognition

Walter Freeman suggests that those aspects should be described as *embodied cognition*, in which the body serves as the principal tool of the brain for cognitive development and the acquisition of knowledge, and as *situated cognition*, in which the structure of the environment determines, and is determined by, the actions of individuals seeking knowledge. Behavior is proactive, not reactive. This model is central to a variety of cognitive systems such as pragmatism, existentialism, Piagetian developmental psychology, Gestaltism, its derivative, ecological psychology, and embodied cognition. In these modern views, information is implicit in the objects that the brain is seeking. The search is initiated when the brain creates a goal with a need for information to realize that goal, and the brain directs the sense organs to find the information in the world, using the cognitive map in the medial temporal lobe to direct the search. At the same time the brain prepares the sensory circuits by tuning them with copies of the motor commands called 'corollary discharges' (Freeman, 2001; Nunez, 2000) that selectively sensitize the cortices to the desired input. The information carried by objects is detected by the senses. They send it to the sensory cortices where it is extracted by resonances in the neural circuits that are tuned just before the search takes place. These actions and the prior tuning constitute the exercise of foresight and selective attention. They limit the entry to the desired information, not whatever forms or energies from irrelevant objects happen to enter the sense organs or be put there by a naive experimenter. A key type of information selected by the resonances corresponds to the uses to which the objects are to be put, their *affordances* (Gibson, 1979).

This also is the domain of recurrent neural networks using unsupervised learning, which is the blind probing for structure that extracts information by gradient descent in high dimensional measurement spaces. Thus the manner of acquisition of knowledge is by *thrusting the body into the world*, from which our word 'intention' has come from the Latin *intendere* = 'stretching forth'. The thrust initiates the action-perception cycle, which is followed by the changes through which the self learns about the world, by assimilation (from the Latin *adaequatio* = adequacy) of the self to the world. There is no transfer of information across the senses into the brain, but instead the creation of information within the brain under the existing constraints of the brain and body.

In this respect cognition is related to digestion, which protects the integrity of the immunological self by breaking all forms of foodstuffs into elementary ions and molecules, which are absorbed and built into complex macromolecules, each now bearing the immunological signature of the individual self. Similarly, events and objects in the world are broken into sheets of action potentials like pinpoints of light, the *raw sense data* of analytic philosophers and the *phantasms* of Thomists, and new forms emerge through constructions by the chaotic dynamics in sensory cortices.

The explanation for this manner of function of both the neural and the digestive systems is essentially the same: the world is infinitely complex and the self can only know and incorporate what the brain makes within itself. This is why neurobiologists using passive neural networks cannot solve the figure-ground problem, why linguists cannot do machine translation, why philosophers cannot solve the symbol grounding problem, why cognitive scientists cannot surmount the limitations of expert systems, and why engineers cannot yet build autonomous robots capable of operating in unstructured environments. The unbounded complexity of the world defeats those classic Platonic and Aristotelian approaches.

3.10.1 Landscapes

Every sensory cortex maintains a landscape of chaotic attractors that correspond to the perceptual categories that the subject is capable of discriminating. *When the subject attends to an expected event, the landscape is created, and a known stimulus gives access to the proper basin. After the dynamics converges to the selected attractor, the specific details of the stimulus are removed. The raw sense data by which category selection is performed are discarded after they are no longer useful.* These properties of chaos are most in evidence when a hypothesis is disproved by testing, because the resulting stimulus is novel. Upon this failure, the prediction is changed by creative activity in the limbic system and disseminated by corollary discharges, so that a new hypothesis is tested. The process is repeated by the creation of new attractors by trial and error, until a reward is forthcoming and a hypothesis is proved.

The changes generated by chaotic dynamics are in the form of the modification of numerous synapses that interconnect the neurons in many parts of the brain. When the process finally succeeds, the changes bring about assimilation not by the incorporation of the forms or the information offered by the world but, instead, by a creative reshaping of the brain as well as the body that facilitates continuing interactions of the self with its world, in so far as that world is accessible to the brain and body. Learning to dance, to play the violin, or to play tennis requires changes in the body as extensively as in the synapses of the brain.

In this view consciousness comes only after an action has been initiated, not before, so that the action is perceived by the actor as a cause, and the sensory consequences as effects. This self-awareness of the intent-action-perception-assimilation cycle is the basis for the concept of causation and for the critical importance of the time lag between effect and its necessarily prior cause. Studies by Piaget have shown that this association of cause and effect is laid down in the somatomotor phase of development, when an infant is learning to control its body. How and why the experience of awareness comes soon after in the way that happens are matters for speculation, but its phenomenology gives clear indication of both the powers and the limitations of this remarkable process. Its temporal range is enormous, but its momentary content is sparse. Our actions are influenced by our entire life history, but we are conscious of only minute, intermittent fragments in sequential moments of thought. There is compelling clinical and physiological evidence that the medial temporal lobe is essential for the construction of the life history, which gives the wholeness of intentional structure, the self, personhood, or, more popularly, personality. However, that brain part is not necessary for consciousness, nor is the hippocampus a site of long term memory storage. On the contrary, the dynamic processes that comprise intentionality in action are non-local, and likewise the personality to which it gives rise. Furthermore, most of intention is inaccessible to consciousness at any one moment, not so much that access is forbidden, rather that the sheer massiveness of the dynamic state space precludes squeezing the entire past through the eye of a needle. Reification of intentionality as the unconscious appears to reduce this living body of experience inappropriately to a static warehouse of latent memories with locked doors.

New developments in the sciences of nonlinear dynamics, complexity and chaos indicate that this conflict is a pseudo-problem (Freeman, 2001). No human action is entirely determined by genes, world or self but instead by an ever-shifting balance of interactions among the three formants. The challenge for neuroscientists is to analyze the brain dynamics by which goals emerge in the form of imagined future states, which lead to neural patterns of selective attention, pre-configuration of expected input, and actions that are intended to bring into conformance the flow of sensory input with the desired state. How do we come to dream of that which should be but is not, and then act to realize the dream?

3.10.2 Image search

An important hypothesis in neurophysiology supposes that sensory information is encoded in spatial patterns of neural activity. A test of this hypothesis in rabbits, for example, reveals that the spatial pattern of encephalographic (EEG) activity of the olfactory bulb depends less on odor than on expectation of odor. Analysis of the structure and nonlinear dynamics of the olfactory system suggests that the EEG manifests a neural activity pattern that is an active filter corresponding to what ethologists call a *chemical search image*. This first glimpse at the physical aspect of what may be a primitive form of mental image allows us to infer some of its properties. It is self-organized; input destabilizes the bulb, and the event runs a course that is shaped by the initial conditions, the sensory input, the chemical state of the bulb from centrifugal input, and the record of past input embedded in synaptic connections. The event is analog, but not an analog of input. Its intensity pattern exists in two dimensions but it is not a picture. Its bases in feedback and resonance make it connotative and ascriptive, not denotative or descriptive. The *neural-mental image* is an operator, not an operand. It gathers neurons by the millions into coherent activity and creates information, which it transmits to the next stage in the brain, thereby helping to shape behavior. It cannot be understood without reference to both what the subject does and what is done to it. Its function in the bulb is to regulate olfactory input with minimal and non-specific centrifugal control.

The relative invariance of the EEG spatial pattern gives an essential clue to the behavioral significance of the neural mechanism of the bulb. First, the synaptic template can provide for an active filter that selectively enhances the sensitivity of the bulb to expected odors without impairing its responsiveness to unattended odors. The suppression of responses to unwanted odors can also occur but by other neural mechanisms that effect habituation. By these means the bulb can abstract selected input. Second, the bulbar activity that is governed by the template can be regarded as a generalization over particular combinations of receptor input, such that there is a characteristic output for every presentation of a given odor CS, whether or not the particular combinations occurred during the training period. In psychological terms the template can provide for an invariant response over a class of equivalent stimuli. Third, the spatial pattern of bulbar activity can serve to represent an expected CS even if the particular odor is not present. For these reasons Freeman has called the activity pattern the manifestation of an olfactory *search image* for the odor CS (Freeman, 1981), following the practice of ethologists who have established behavioral evidence for the existence of search images in numerous animal species.

The search image may be said to dominate bulbar output but not to the exclusion of other output relating to other odors. For example, an unexpected odor previously given significance leads to the orienting response and an activation of a dormant template. A persisting novel odor if reinforced also induces the orienting response and formation of a new template. It is in this condition wherein the novel input is still "undefined" for the subject that the

EEG odor-no odor differences are detected. One may indeed ask how far into the nervous system raw sensory input can pass; the answer is not yet known. If this line of reasoning is correct, then these EEG contour plots reveal our first glimpse of the physical aspect of a mental event. Even though we cannot know whether rabbits introspect, we could call it a *mental image* partly because neural and mental images are two sides of the same coin, and partly because we want to attract the attention of cognitive psychologists, who believe that such *images* or *representations* must exist, and who now vigorously debate their nature. All agree that they are products of the brain and that they form under the physical and chemical constraints that govern the brain. This primitive example allows us to speculate about physical properties of brain images.

3.10.3 Analog vs digital

Clearly the event under observation is analog as opposed to digital. It consists in the cooperative activity of indefinitely large numbers of neurons and their continuously variable dendritic currents, axonal pulse densities, and concentrations of chemical transmitters and modulators. Its local manifestation occupies several cubic millimeters of tissue without sharp boundaries. *A proper language for description is by integro-differential equations of classical physics and not Boolean algebra as applied to logical switching networks.*

Yet it is not an analog of that which it represents. An odorant and the neurons of the receptor layer and bulb are made of the same substance that obeys the same laws of physics and chemistry, but there is no other way in which the physicochemical properties of the odorous molecules resemble the neural activity that follows reception. *Is it a picture or a description?*

During normal respiration there is a sequence of oscillatory bursts in the two surface dimensions of the bulb; in this respect the image more closely resembles the frames in a motion picture than it does a one-dimensional tape recording of a spoken sentence. Yet it is not a picture, which is a static distribution of pigments on a surface, nor does it represent literally the olfactory ambience at a given time. The pathways for lateral interaction within the bulb and cortex that shape the patterns of cooperative activity within them and the resonance properties between them suggest that *this event is not denotative and descriptional (determined by forward gains) but is connotative and associational (determined by feedback gains).* The traces of an odor that are stored in synaptic memory (the engram) are not unitary, serially ordered, or episodic; they are the cumulative record of the co-variances that were transiently induced among central neurons by receptor input, during periods of learning in which a certain category of input was received.

The olfactory burst is an ephemeral activity for which the pattern is subject to continual variation on successive inspirations, but it is not epiphenomenal. It is a dynamic process that draws its shape from the olfactory environment, from the residues of past environments embedded in the synaptic connections, and from the chemical state of the bulb controlled by centrifugal input from the brain. It is an operator that sweeps up neurons by the millions into coherent and cooperative activity, thereby creating information. For the subject it is the information, because it shapes the output of the bulb. It provides a "form" for Merleau-Ponty's *structure of behavior* (Merleau-Ponty, 1960).

3.10.4 Observers

Perhaps the strongest objection to use of the concept of mental image could be an implicit requirement for a reader or viewer of the image. The physiological properties of the image suggest that there is no such requirement. By extrapolating the interactive properties of the bulb and cortex to other parts of the limbic system, such as the amygdalo-striatum and hippocampus that have similar structures and dynamics, we can conceive of the activation or creation of images of expectant motor activity (Freeman, 1981).

It should not be lost to you that the search-image hypothesis is itself a search image. EEG waves do not carry labels to identify them as images; that interpretation stems from the context of the anatomy and physiology of the bulb, the history of behavioral testing, and the mathematical model that describes the neural dynamics, in which the key feature is the sigmoid curve. It postulates that *information processing* is a highly interactive neural process that assigns incoming information to pre-existing categories by abstraction and generalization. It comes close to restating contemporary psychological views on the process of perception, but it emphasizes the creative aspect of that process.

The image we would like to invoke is that *with each breath the rabbit draws it creates a new world view, and if rabbits do that then surely so do we.*

3.11 *Simulacra*

As expressions are the concrete representational vehicles of a Discrete Formal System, so images can be regarded as the concrete representational vehicles of a Continuous Formal System (MacLennan, 1995).

An image displays a continuous pattern of bounded variation over a bounded continuum. The well-formedness (syntactic correctness) of an image depends continuously on its pattern of variation. All matters of syntax are continuous; that is, they are matters of degree.

Many of what we call images here are images in Peirce's sense, since they correspond to their referents in a direct way; other representations, which we call images (such as Fourier or wavelet transforms), might be better classified as diagrams, since their relationships to their referents are more abstract. Images are also found in the brain: consider the continuous variation of electrochemical activity over the neuropile. (Even if we take discrete synapses to be the sites of activity, the distribution is practically continuous, since there are approximately a billion (10^9) synapses per mm_2 for cortex.) In general, any continuous quantity extended over space, time or other dimensions (e.g. frequency, energy) may serve as the substrate for an image. A reasonable standard for physically realizable images is to require that they be finite energy functions: i.e. functions with a finite

$$\Lambda_2 \text{ norm: } \int\Omega\phi^2(x)dx < \infty.$$

Another issue relevant to the physical production of images is their dimension (the degrees of freedom of their extent). Mathematically, images can extend over any number of dimensions, but practically they are limited to a few dimensions.

Physical images can extend over time, over at most three space dimensions, or over various other physical dimensions (frequency, energy, orientation, color, etc.), either singly or in combination. Analogously, although in principle the formulas of a DFS may be of any dimension, in practice they are one-dimensional strings or occasionally two-dimensional arrays of characters.

The issue of syntactic correctness is one of interpretability. In traditional logic only the syntactically correct expressions are interpretable. *In continuous logic, images are interpretable to the extent that they are well-formed.* Continuity does not preclude there being some images that are entirely non-interpretable and others that are entirely interpretable, but there must also be intermediate images with intermediate degrees of interpretability. So is, also for well-formedness. The meaning of an image depends continuously on the form of the image, and an image has a meaning to the degree that it is well-formed (syntactically correct).

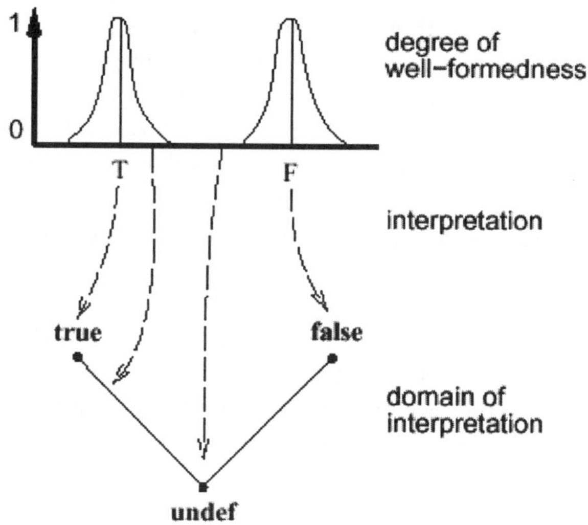

Inherent fuzziness of syntactic well-formedness (MacLennan, 1995).

There is an inherent fuzziness of syntactic well-formedness. Just as the degree of well-formedness decreases continuously from 1 to 0, so also the interpretation must vary continuously from a defined interpretation to an undefined interpretation.

The interpretation of an image is defined:

$$\Im(\phi) = \mu_T(\phi)\, true + \mu_F(\phi)\, false + [1 - \mu_T(\phi) - \mu_F(\phi)]\, undef$$

In a CFS the computational process, by which one image is derived from another, is continuous; that is, the computation comprises a continuous process of continuously changing direction. *At each point the infinitesimal change effected by the process is a continuous function of the form (shape) of the image, and the result of the change is a continuous function of the image at that time.*

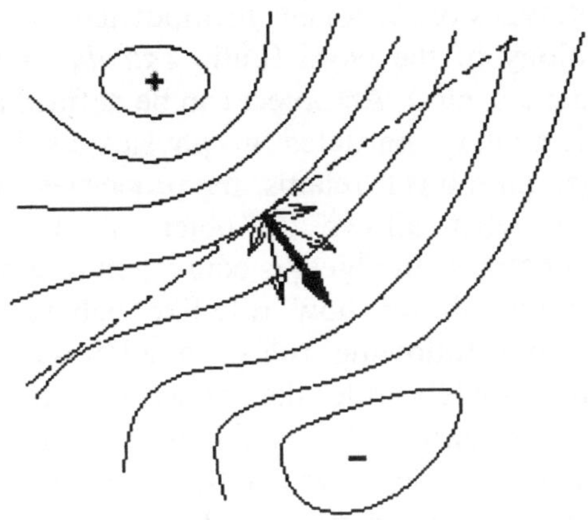

Nondeterministic computation by descent on potential surface (MacLennan, 1995)

The facility of change in a given direction is shown by the length of the arrow (which is proportional to the cosine of its angle with the negative gradient. The dotted line separates impossible changes to the left from possible changes to the right. A continuous generative grammar is the guiding image of a nondeterministic computational process capable of generating, with varying facility, terminal images from a fixed starting image. There will be a *continuum* between images generated by the grammar and those not so generated, so *well-formedness is a matter of degree.*

3.11.1 Discrete emergence

How do (approximately) discrete formal systems *emerge* from continuous representations? The answer to this question is a step towards *understanding the origin of apparently discrete faculties, such as language, from continuous sensory-motor processes.*

An important part of this process is the perception of *objects*; indeed, objects are the prototypes of the tokens manipulated by calculi, as we can see even in the etymology of the word (Latin *calculus* = small stone used in counting, voting, games, etc.). An object can be defined as a *stable bundle of properties* (cf. elementary particles in physics as bundles of quantum numbers). That is, an object retains its properties when its context is changed, which is what allows an object to be separated from its background. In other terms, an object is equivalent to *a set of invariances*.

Connectionism has shown how rule-like behavior can emerge from processes that are not following rules. In addition, it demonstrates the possibility of approximately rule-like behavior, which exhibits rule-like behavior when it is appropriate, but is able to bend or adapt the rules when that is more effective. In this way we can begin to understand and imitate the exible rule-like behavior of the natural world.

We have shown that the concept of a continuous formal system (or *simulacrum*) is analogous to a discrete formal system (or *calculus*), but that the former is continuous where the latter is definite. Nevertheless, the theory of CFSs forces revisions in our notions of syntax, semantics, computation, program, grammar and rule. It promises a new theory of representations in artificial intelligence, cognitive science, linguistics and philosophy, which can better address the emergence of discrete entities, such as objects and rules, from the underlying continuous processes. In particular, we have considered how a multi-resolution representation of change can aid the detection of discrete structures. Continuous formal systems will provide a theoretical framework for understanding emergent and adaptive information processing in all its manifestations.

Chapter 4: Couplings

Experience is remoulding us at every moment:
whilst we think, our brain changes.
William James

4.1 Antecedents

According to legend, Galileo discovered the principle of the pendulum while attending mass at the Duomo (cathedral) of Pisa, Italy. A few years later, in 1673, Huygens published *Horologium Oscillatorium*, in which he discussed the pendulum clock of his design and the theory on which it was based. In the winter of 1656-7, he had invented the pendulum clock. He also reported on his observation of *synchronization of two pendulum clocks on the same wall*: the *tiny vibrations of the wall*, as he well understood, *had forced the clocks to synchronize.*

Synchronization means adjustment of frequencies of periodic self-sustained oscillators due to weak interaction. This effect of the *coupling* of the two pendulum systems (also referred to as *phase locking* or *frequency entrainment*) is well studied and has a lot of practical applications. Systematic study of this phenomenon, experimental as well as theoretical, was started by Edward Appleton and Balthasar van der Pol. They showed that the frequency of a triode generator can be *entrained, or synchronized*, by

a weak external signal with slightly different frequency. These studies were of high practical importance because such generators became basic elements of radio communication systems. Mutual synchronization of two weakly nonlinear oscillators was analytically treated by Mayer and Gaponov. An important step was done by Stratonovich who developed a theory of external synchronization of a weakly nonlinear oscillator in the presence of random noise. Recent important developments are due to the Pecora and Kurths groups (see next ch. and Pikovsky, 2001).

4.2 Biological rhythms

The *natural rhythms of life*, in bodies, minds and populations, can be considered in their functioning as similar to pendulum clocks. When we are dealing with more than one in the same system, we should consider the possibility of synchronization. Why are there Biological Clocks and Rhythms?

All biological clocks are adaptations to life on a rotating world. Biological clocks allow organisms to assess and respond to the oscillating environmental rhythms that result from the Earth's movements via the generation of biological rhythms.

What are *Environmental Rhythms*? Environmental rhythms are things like day and night and the rise and fall of the tides. They arise because of the following synchronizing influences:

1) The Earth's Spin. As a result of this we have the solar day (approximately 24 hours). This is responsible for the generation of other 24 hour rhythms, one of the most significant of these being the light-dark cycle. As we shall see later the light-dark cycle is influential in the establishment and maintenance of many other rhythms e.g. locomotor activity patterns.

2) Movement of the Earth around the Sun. Results in the annual period known as the year which is 365 days, 6 hours, 9 minutes and 10 seconds long. The Earth has an elliptical orbit (it is tilted on its axis at an angle of 23.5°) and this causes the generation of the seasons as the Earth tilts towards and away from the sun. The seasons have important implications to organisms for a number of reasons: they dictate when animals can breed and when some must enter hibernation.

3) Movement of the Moon around the Earth. This generates the lunar month (24.53 days) and the lunar day (24.8 hours). Many organisms have rhythms that correspond to these periods.

The gravitational pull of the moon also causes the tides: there are normally 2 tides per lunar day i.e. every 12.4 hours. For organisms living in the tidal zone of the seashore this has a number of implications (it will govern when they can feed for example). The combined effects of the sun and moon's pull causes neap and spring tides which occur every 14.79 days, and for some marine organisms these are important times in their life cycle: perhaps controlling when they can reproduce.

Daily rhythms in plants and animals have been noticed since early times. As early as the fourth century B.C., Alexander the Great's scribe, Androsthenes, noted that the leaves of certain trees opened during the day and closed at night.

Many years later, the famous taxonomist Carolus Linnaeus discovered that the petals of many flower species opened and closed at regular times. He even created a garden with flowers which opened at various times so that he could tell the time of day by looking in his garden.

In 1729, French astronomer Jean Jacques d'Ortous deMairan conducted the first known experiment on biological rhythms. He noticed that his *heliotrope* plant's leaves opened during the day and folded at night. When he put the plant in total darkness, the rhythm continued, and was not disrupted by the absence of daylight. Although he was interested in this botanical phenomenon, deMairan pursued astronomy instead.

Nothing exciting happened for many years...until the early 1900's, when Karl von Frisch observed that bees visited flowers only at specific times. He and Ingeborg Beling trained bees to visit a nectar feeding station between 4 and 6 PM. The bees did not visit at other times, and they still visited even when the nectar was removed. When outside cues such as light was removed in laboratory trials, the bees still fed at prescribed times. Although von Frisch did not know it, the bees were operating on an *internal clock*.

It wasn't until the 1950's that Gustav Kramer and Klaus Hoffmann proved the existence of a biological clock. With an ingenious apparatus, Kramer demonstrated that starlings used the sun as a compass to migrate even though the sun itself moves throughout the day. That is, the bird's internal clock reorients it in the direction of the moving sun. Hoffmann showed that the clock persisted in dim light and thus is endogenous to the animal. He showed that the animal's clock was synchronized to local time by the influence of the local environment (Cornelissen, 1994).

4.2.1 Chronobiology

Since the early experiments, *chronobiology*, the study of biological rhythms, has become established as an interdisciplinary field within biology. Most chronobiologists study *circadian rhythms*, endogenous cycles of behavior or biological activity with a period of about 24 hours. In the example, the human sleep-wake cycle has a period of 1 day, or goes through 1 complete cycle in a day. Circadian rhythms, like the sleep-wake cycles, are generated by an internal clock that is *synchronized to light-dark cycles* in the environment and *other daily cues*. An *entraining agent* (for example, exposure to bright light) can cause a phase shift whereby the activity is started earlier or later in the day.

Like a watch, the circadian clock must be synchronized to local time. For example, animals kept in total darkness will show a free running rhythm that is independent of the local time. A circadian clock is most useful, however, when it is set to local time; the animal must be in sync with its prey, pollinators, and other members of its social group in order to survive. In mammals, the light-dark cycle is a major synchronizer or entraining agent for circadian rhythms. The clock in humans is located in the *supra-chiasmatic nucleus* (SCN), a distinct group of cells found within the *hypothalamus*. The SCN is only one part of the mechanism by which the time is kept. There are light receptors found in the retina which have a pathway, called the retino-hypothalamic tract, leading to the CNS. The pineal gland is a pea-like structure found behind the hypothalamus in humans. The pineal gland receives information indirectly from the CNS. It appears that the CNS takes the information on day length from the retina, interprets it, and passes it on to the pineal gland, which secretes the hormone melatonin in response to this message. Night-time causes melatonin secretion to rise, while daylight inhibits it. Even when light cues are absent, melatonin is still released in a cyclic manner; yet if the SCN is destroyed, circadian rhythms disappear entirely.

The CNS also plays a role in the circadian system by triggering a *neuroendocrine response* in the hypothalamus, which then acts on the pituitary. This last pathway profoundly influences other parts of the body, including the endocrine, immune, cardiovascular, and urinary systems. Rhythms in most of these systems have a simple waveform similar to that of body temperature, which is highest in the early evening and lowest right before waking in the morning. A plot of urine volume, for example, shows a cyclic pattern very similar to that of the body's temperature.

Some of the hormones thought to be influenced by the circadian system are growth hormone, prolactin, thyrotropin and testosterone. An endocrine event clearly under the influence of the clock is the release of the hormone ACTH. The CNS triggers the hypothalamus, which activates the anterior pituitary to release ACTH, causing the adrenal glands to release cortisol and aldosterone. A plot of cortisol concentration in blood plasma shows a characteristic peak in the very early morning (around 6 AM) with a trough right before bedtime. Similarly, aldosterone level is constantly high throughout the night and low throughout the day.

Activity of the *immune system*, as represented by the number of lymphocytes, also seems to peak in the late evening and is lowest a few hours after the cortisol peak in the morning.

The timing of the endocrine and immune systems are clearly intermeshed. It is known that aldosterone and cortisol suppress the immune system, while melatonin appears to enhance it (Cornelissen, 1994).

4.2.2 Darwin

Charles Darwin is probably most famous for many other things. In his later years, he became fascinated with *plant movements while trying to find an evolutionary relationship among them.* He did hundreds of experiments to keep track of the movements of the different varieties of plant leaves. After experimenting, Darwin concluded that the plants were moving their leaves so as to expose the smallest possible leaf surface to the night temperatures. His book, "The Power of Movements in Plants" (Darwin, 1977) details the years of work. In the years following, scientists would debate whether the rhythm arose from forces external to the plant or was endogenous.

The term *photo-periodism* was used to designate the response of the organisms to relative length of day and night. The ability to sense day length is an important ability for plants so that they grow, reproduce and develop during favorable time of the year. The changing times of dawn and dusk contain seasonal information as well as time of day information so that the organisms have, in effect, an internal clock and calendar. In plants a photoperiodic clock not only controls flowering, but also induction and termination of dormancy in buds and bulbs, seed germination, and daily rhythms such as leaf movements, petal movements, and nectar secretion.

4.2.3 Gene rhythms

Recent experiments with plants have developed an interesting technique to measure rhythms in Arabodopsis thaliana. The researchers chose this plant for many reasons including its small size, short life cycle, and number of chromosomes (n=5). As a way to measure *rhythmic gene expression* in vivo in these plants, they transplanted firefly luciferase gene (which is responsible for the insect's glow) into the plants. The plants exhibited rhythms in bio-luminescence when their CAB genes (CAB is a light harvesting protein) were being "turned on" rhythmically. In other words, the luciferase gene was used as an intra-cellular marker to detect the plants' own gene activity rhythms.

As every parent knows, infants are seldom on the same sleep-wake cycle as the rest of the family. In human infants, the sleep-wake cycle is initially nearly random. At about 6 weeks of age, the circadian rhythms begin; by about 16 weeks, the baby is entrained. Do infants, in fact, have an endogenous circadian sleep-wake rhythm, or do social cues (parents' schedules) directly control their behavior?

Under normal conditions, the often complex social environment in which most humans live presents many *Zeitgebers*, time markers, such as our work schedules or daily caffeine consumption. These factors work with or against our entrainment to the natural light-dark cycle.

Early sleep-wake experimenters eliminated all external cues, such as light and sound, to understand the human sleep-wake cycle. To understand the free running sleep-wake cycle, one of these researchers, Jurgen Aschoff, built an isolation facility known *Tier Bunker*. Starting in 1961, test subjects stayed in the bunker for 3 or 4 weeks and revealed a sleep-wake cycle of about 25 hours. (Recent research suggests that the human cycle may be closer to 24 hours). They could turn light on and off at will, but no external cues could reach them. Under these controlled conditions, all subjects displayed clear circadian rhythms, and experimental changes in room temperature seemed to have no effect on their cycle.

4.3 Dynamical diseases

Classically, physicians have long recognized the importance of considering the *temporal dimension of illness* for arriving at a diagnosis and deriving treatment strategies.

Diseases can be distinguished, in a temporal dimension, by their onset (acute versus sub-acute) and by their subsequent clinical course (self-limiting, relapsing-remitting, cyclic, chronic-progressive). Recognition of the temporal rhythms, combined with the knowledge that certain temporal rhythms respond best to certain treatment strategies, often provides a basis for therapy.

The concept of dynamical disease was first proposed by Mackey and Glass (Glass, 1988). Dynamical diseases arise because of abnormalities in the underlying physiological control mechanisms. The hallmark of a dynamical disease is *a sudden qualitative change in the temporal pattern of physiological variables*. The significance of identifying a dynamical disease is that it should be possible to *develop therapeutic strategies based on our understanding of dynamics combined with manipulation of the physiological parameters back into normal ranges*.

The extraordinary range of dynamics found in human functioning it is one of the fascinating and challenging features in the study of this field. Collecting data is one of the first steps, and its methodology is half-way between science and craft. Detecting *time textures* in different media is the first step to obtain materials to study using the various tools of time series analysis.

The identification of some deterministic dynamics could allow the design of stimulation strategies and procedures based on nonlinear dynamics (Kupfer, 1988; Gooddy, 1988).

4.3.1 Neuroscience

Various studies in the field of physiopathology propose a viewpoint on illness as a change in a physiologically intact system that operates within a defined range of control parameters which can, at times, cause it to take anomalous dynamics. This is related to the *sensitivity to initial conditions of nonlinear-dynamics*: the development of a disease may be associated with passages from one dynamic regimen to another. These *bifurcations* have also been observed in neurology especially in studies on epilepsy. The *dynamic approach* has already produced some clinical results towards the prevention and treatment of arrhythmias and sudden cardiac death. Studies on heart dynamics now represent a paradigm for other studies on dynamical diseases (Goldberger, 2002; Goldberger, 2002a; Goldberger, 1992; West, 1999 & 2001).

Based on the analysis of temporal variations in cardiac frequency (R-R space in the ECG), cardiology studies have shown how the healthy heart has a variable frequency, organized by strange attractors. Old or sick hearts, on the other hand, are more "ordered" in the sense that the variability of their parameters is much more restricted: *they occupy a smaller area in the phase space*. This implies a reduced elasticity of the system, which is more vulnerable by various stressors, and thus a tendency toward developing disease. Thus, physiologists Goldberger and West came at conclusions: “healthy systems don't want homeostasis, they want chaos”. They explained that, if we accept that our anatomy is fractal, only a small step further is needed to recognize that even the function of physiological systems is based on fractal geometry: form and function are correlated.

The dynamical approach, based on systemic complexity as a health factor, has been applied to the study of various degenerative processes and decay: Parkinson's, epilepsy, senescence etc. The results of these studies imply a revision of Cannon's homeostasis in that *a biological system is found to be powerful, efficient and adaptable if it is complex, whereas disease and decay have in common the simplification or disorganization of system dynamics. There is an optimal, healthy, status of the system functioning that is tuned at the dynamical state of Self Organized Criticality.* New proposals for the experimental treatment of degenerative diseases are based on nonlinear control and anti-control.

4.3.2 Mood disorders

Chrono-psychobiology concurred to the dynamical study of mood disturbances and the related disorganization in psychobiological rhythms.

Each rhythm is considered in physical terms as an oscillator, and studied as such in its couplings or de-couplings with other rhythms in the same system of reference. Each rhythm has its own *Zeitgeber*, a metronome indicating tempos and synchronizations. It also has *Zeitstoerers*: psychosocial or physical factors that reorganize rhythms.

The psychobiological perspective established by Myron Hofer identifies life events and psychosocial conditions as potential *Zeitstoerers*. Interpersonal relations and social needs can be recognized as powerful synchronizers for biological rhythms.

Several studies are focused on cycles in mood disorders and their fluctuations in time. These studies demonstrated the existence of a *1/f type time series* in mood variations, according to the so-called *Self-Organized Criticality*, SOC. The combination of a minimal dynamic stability and a spatial self-similarity produce a distribution in time series according to a power-law: relative predictability is the result.

In manic-depressive subjects, transitions from one state to another may be induced by minimal perturbations and are inescapable for the individual. But, using nonlinear methods, within certain levels of intensity and for a given time period, changes in status become predictable. This limited predictability might be used for therapeutic purposes, as pathological oscillations in mood appear to be much more organized than the normal ones. Possible treatments should thus be directed against the observed loops, to *tune the dynamic complexity* of the manic-depressive subject (Thomasson, 1999; Pezard, 1996 & 2001).

4.3.3 Schizophrenia

Hoffman has suggested "a computer simulation for the study of the schizophrenia-mania dichotomy". He was following hypotheses that manic and schizophrenic thought disorders could be overlapping. Manic subjects can establish their discourse on coherent projects or *Gestalten*: the evident disorganization in their verbal productions would derive from irregular leaps between parallel thought processes.

In schizophrenic subjects, on the other hand, the ability to generate coherent thinking would seem weak. Their thought processes are incomplete and disorganized, resulting in speech lapses (Schmid, 1991; Orsucci, 1996).

An interesting experiment from Hoffman's was starting from a simulated neural system, organized according to associative memory and mass action. During the experiment, it became evident how *overloading the system produced parasitic states*. These isolated dynamics could start equivalents of secondary symptoms in schizophrenia (hallucinations, delusions etc.).

Working on a similar perspective, Schiepeck et al. (1990) proposed a simulation of evolutionary profiles in schizophrenia. This simulation was based on nonlinear differential equations, built within a bio-psycho-social framework. They included the so-called diathesis-stress model, interactive evolution, neural plasticity, vulnerability and selective evolution.

The construction of this mathematic model allowed an evolutionary simulation of the patient-system, including manipulations of various control parameters.

Studies like these are included into the framework of neural-dynamics, a new kind of psychodynamics which recognizes the mass action of millions of neurons and billions of connections as the necessary premise for self-organization and plasticity.

This approach tends to turn into obsolete the classical psychopathological distinction: State vs Trait.

Because of the neural plasticity, the temporary passage through a psychotic condition shall leave its imprint on the neural hardware. The global duration of the psychotic state has prognostic and therapeutic importance for the construction of any stable and eventually chronic neural asset.

The necessary reference is to the process of pathological *cortical pruning* that might be produced by the persistence of a psychotic state. The result is a decrease in the number of connections for each single neuron and the consequent reduction in the overall complexity of networks. This change in the network architecture leads to the genesis of new autonomous dynamic attractors acting as parasitic foci in the genesis of secondary symptoms in schizophrenia.

A process as the one described above does not simply imply the increase or decrease in the action of one or more neuro-mediators, but rather speaks for an all-encompassing disorganization and a following re-organization in neuro-cognitive systems.

4.3.4 Alexithymia

We have described a single case study to exemplify some the numerous variables in coupling and decoupling in this neuroscience domain (Orsucci, 1996). This case history shows how an apparently minimal deviation in linguistic patterns can be related to a severe illness, in this way representing the semiotic marker of a systemic disease. One of the most salient features of the subject examined in this clinical history is the disorder in the expression of emotions, connected with a lack of mindedness and preference for attention to the external world. These traits are organized in a specific personality construct that is called *Alexithymia* (meaning literally 'no word for emotion').

The term alexithymia was first coined by Nemiah and Sifneos (Nemiah, 1977; 1996) to define features noted in the so-called "psychosomatic subjects", but the term's definition was expanded to a personality construct that can be considered a paradigm for a cognitive dynamic disorder.

The classical alexithymic traits are:

1) difficulty in feeling, recognizing and finding appropriate words to describe feelings;

2) poor fantasy life and elaborate description of trivial details;

3) poor dream life;

4) interpersonal relationships in balance between strong dependency or avoidance;

5) life-style centered on action;

6) the interviewer finds the subject boring, dull or void.

The prevalence of alexithymic traits has been found in several conditions, such as: drug addiction, somatoform disorders, posttraumatic stress disorder, chronic pain, obesity and anorexia, major affective disorders, masked depression, sexual perversions, sport addiction, chronic degenerative diseases and long hospitalizations.

The idea that a kind of variation in brain organization is linked to alexithymia is supported by the demonstration among comissurotomy patients of scoring in tests higher than in control groups. It is demonstrated as a deficit in the inter-hemispheric transfer in alexithymic subjects with posttraumatic stress disorder, and also in children and adults with evidence of lesions in the right hemisphere. An association between alexithymia and left cerebral lateralization has been found in a sample of young adults with intact brains.

Antecedents of the alexithymia construct are the works of Marty and de M'Uzan who proposed the concept of *pensée operatoire* to define the lack of imagination, the externally oriented thinking of psychosomatic subjects (Marty & de M'Uzan 1978).

This definition of *pensée operatoire* is obviously, if not explicitly, derived from the works of Jean Piaget. He was the first to define this cognitive asset with its specific logical features and the first to propose a specific phase in human cognitive development. Following his observations these are the main features of *operative thinking*:

1) complex cognitive structures allow sophisticated logical operations, but there is a lack of abstraction from physical objects (Piaget lists nine kinds of logical operations available at this stage);

2) capacity to understand differences between experienced events and possible events is reduced to real experiences;

3) forma l logic and the capacity to define conceptual hierarchies are not used;

4) lack of capacity to assume a meta-position or to combine different points of view;

5) difficulties in abstraction, and breaking of links between what is real and what is possible

6) use of irony, play and creative thinking strange or difficult.

The norma l development of this stage in the evolution of thinking is situated by Piaget in the first adolescence. Some authors tried to define integration between Piaget's studies on cognitive development and the psychoanalytical studies on the expression of emotions. They conceptualized a sequence of stages from simple awareness of undifferentiated bodily sensations to an awareness of complex blends of feelings and the capacity to appreciate the emotional experience of others (*empathy*). There is clinical evidence that shows that people who could be placed (by regression or arrest in development) in the lower stages of this developmental line have *a rigid asset in the regulation of emotional and psycho-physiological balances*. If these subjects are going trough stress they are prone to regulate their emotional states manipulating other people, consuming alcohol, drugs and food in excess. They also tend to experience some *strange physical distress*, caused by instability of the autonomous nervous system, and undergo a somatic disease or a casualty (car or sport accidents etc.).

4.4 Systemic health

The description we have given of alexithymia renders the main aspects of this psycho-neural asset as a *dynamical or systemic disease*. It is evident how some connections (between language, imagination and emotion) are truncated and some others are reinforced or modified. Some new and paradoxical systemic behaviors are established.

Alexithymia can be easily included between systemic disorders, like senescence, Parkinson, epilepsy, diabetes, sudden cardiac death etc.

The teaching we might receive from this studies could be valid for all the domain of complexity theory in the life sciences: there is a range of functioning of a system in which its complexity is optimal. This allows flexibility in its functioning due to a variety of possible behaviors permitted by its dynamic variability. If this variability is restricted or enhanced beyond these optimality limits the system "gets sick", and it risks a dynamical death either by collapse (periodic or fixed dynamics) or breaking of boundaries. In the meantime, on the path towards dynamic death, there can be chronic evolutions leading to unstable regimes based on new pathological attractors. *A biological system is found to be powerful, efficient and adaptable if it is complex, whereas disease and decay have in common the simplification or disorganization of systemic dynamics*.

A systemic therapy is a physical concept related to the procedures aimed to restore dynamical optimality. These procedures imply various degrees of control, synchronization and co-evolution towards stimulating the appropriate rhythms of disorder or order to the sick system.

4.5 Psychobiological couplings

The research conducted by Emde and Spitz (Emde, 1965; Spitz, 1969) on institutionalized infants and their *anaclitic depression* is a basic work. It was followed, during the fifties and sixties, by Engel and Schmale (1972) who observed a higher risk of disease in individuals who were unable to cope with separations or object loss. These subjects developed a "giving up given up complex" whose biological implications have been widely proved and explored in recent years: profound dis-regulations in the endocrine, immune and neural mediatory systems.

Following Bowlby's observations (Bowlby, 1978 & 1982) on attachment behavior, psychosomaticists mainly studied the *mother-infant bond: a strong coupling organization regulated mainly by its "inner" emotional signals*. The strength of this coupling make it so protected and differentiated from influences coming from the environment, to make it an ideal sample for the study of psychobiological couplings. In a series of studies on rodents and primates, Myron Hofer and other developmental biologists demonstrated many hidden regulatory mechanisms (Hofer, 1984; Hofer, 1994). These hidden multiple and pre- and intra-emotional factors act on different sensorial channels: nutritional, olfactory, tactile, thermal, visual and vestibular.

For example, the importance of bodily contact and tactile stimulation was demonstrated by the finding that a decrease in the levels of growth hormone, GH, in separated rat pups can be prevented by stroking their skins with a brush. Also, a 30% reduction of heart rate following separation could be prevented by providing the pups with a feeding. Other, the body temperature of infant rats, which is determined largely by the body temperature of the mother, has been shown to regulate levels of brain peptides, nucleic acids and neuro-amines. All of them are reduced, if the young rats are prematurely separated from their mother.

The olfactory stimuli are also involved in the regulation of crucial aspects: for example infant rats are unable to locate the nipple in absence of a pheromone secreted from the mother's areolar glands, whose secretion is stimulated by the suckling (interesting learning recurrence).

There is evidence of a multiplicity of regulatory mechanisms. Most of them remain *hidden* from a passive third observer, outside of the "nursing-couple", and can only be discovered in experimental contexts. Also in humans these kinds of regulatory mechanisms are operated on different levels than overt emotional expression and language. They are supposed to continue in adult life, but their functioning becomes intermingled with other emotional, cognitive and social factors. They, anyway, play a basic role in growth and health.

The works by Margaret Mahler (Mahler, 1968; Mahler, 1975) on the vicissitudes of human symbiosis and the contributions of Bowlby, Winnicott, Searles, Gaddini and Kohut can be a general framework for these findings. The interactions between the self and its self-objects (also as *transitional objects* or *sensation objects*) regulate the psychosomatic balance.

Synchronizations and de-synchronizations of biological rhythms in individuals and groups can be triggered by a variety of interpersonal relationships and social demands. A common experience of this kind of hidden psychobiological mechanisms in everyday life is related to the well known synchronization of menstrual cycles of women living in the same ambient as, for example, roommates in a college.

A similar, but more intriguing, experience has been told by a student of psychology. She went to live in an apartment where there were living two students. They where single and already synchronized in their menstrual cycle. She remained desynchronized probably because, as she explained, she was emotionally distant from their close friendship, and she had a boy friend. Some months later, her boyfriend had to stay away for a long time due to military service. She, feeling alone, became nearer to her apartment mates, and she was surprised to find synchronized her menstruations with theirs. After several months, when her boyfriend came back in town, she desynchronized again.

The interactions in our everyday life are formed by streams of semiotic entities, and we are unaware of most of them. They act on our minds and bodies. In this way we can come back to the original *Es /Id*, whose name was derived by Groddeck from Nietzsche as "...what there is of not-personal and, we can say, necessary by nature in our being" (Groddeck, 1923).

4.6 Minds in synchronization

Inter-Subject synchronization is what makes possible a dialogue. Intra-Subject synchronization is what makes possible perception and thinking (inner dialogues), in minds and brains. A dialogue is an extended interaction between two systems, and a good example of a natural interpersonal synchronization. A conversation between two people is a dialogue (Have, 1999; Pask, 1976).

Dialogue style varies with the granularity of the interaction.

There are certain characteristics that it is desirable for dialogues to have:

- Syntax: dialogue atoms create meaning by their relationship to one another according to a set of rules;
- Reference: parts of the dialogue can refer to other objects in the context;

- Consistency: the dialogue stays consistent with the context that has been built up;
- Congruence: the dialogue's metaphors are consistent with one another, and with our metaphors from the real world

Mind synchronization is about the coincidence of information. We must synchronize the different streams to make sense of them. We find a *sequential synchronization* when two people have a conversation and only one person talks at a time. When one person is done talking, she gives tacit signs that it is the other's turn. It is impossible for both people to be talking at once; imagine listening, processing what is said, composing a response, and delivering it, all at once.

We must synchronize sequentially with each other in order to have dialogues. Humans have a limited amount of attention, and can only process so much at once. Through synchronization, we multiply our attention between what we are saying and what the other is saying, and make it possible to have a conversation between people.

When we are having a conversation with someone, we follow a certain *protocol* to help manage *turn-taking*. A sharp intake of breath, body language, and the raising of intonation all suggest a passing of speaking rights. This protocol exists to allow us to synchronize with each other and carry on a conversation. We use prompts and triggers to signal that it is the other's turn. It is like the so-called "hand-shaking" of modems. Synchronization eliminates intermixture of information. Since we can only consciously process one thing at a time, we can't process overlapping pieces of information that differ with each other.

By using *sequential and multi-stream synchronization*, we prevent information overlap, and only deal with one stream of information at a time. When streams are synchronized with each other, we understand them as being part of the same object; the streams fuse and we only need to think about one object, and hence once piece of information.

One strategy for dealing with these intermixed streams is to *serialize* them. Each stream gets a turn with our attention so we only need to focus on one thing at once, interacting with only one at a time. In multimedia systems, multiple streams are sometimes correlated such that they appear to be emanating from a single object. This sort of synchronization fuses (merges) the streams together.

4.6.1 Natural vs controlled dialogue

The so-called psychotherapies, in their various forms (individual, group, family or community) are specialized forms of interpersonal interaction in which one human system (often called therapist) tries to influence the functioning of another human system (often called patient or client) to restore a healthy or a better functioning.

This definition, in the asymmetry of roles, unfortunately recalls other definitions used in control theory: master and slave systems. The term patient comes from the Latin *patior,* suffering, that brings with itself meanings of pain and passivity. If we tend to consider the interaction as a form of synchronization and co-evolution (than of control), we should find different terms for the other. We should see the therapeutic partners as forming a "loosely coupled self-organized ecosystem", in which the partners, after synchronization, co-evolve (Palombo, 1999). So we will use the term 'therapeutic partner', P.

We might examine two different examples of synchronization, and co-evolution. Both have been studied using the Recurrence Quantification Analysis, following the methodology explained in Ch. 3.

In the Sample 1 we find a dialogue between two friends: they are talking about "the meaning of life" in a very relaxed way. Each step in the plot represents one turn in the conversation. In the friendly conversation we can see how they start in good synchronization, while they tend to diverge during the last steps of their dialogue

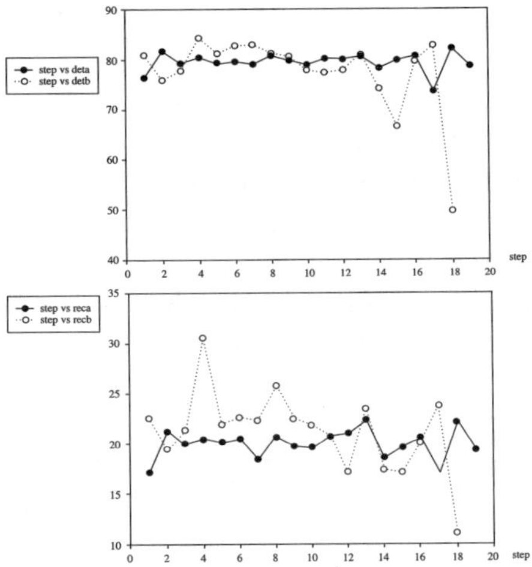

Friends' dialogue (A, black spots, B white spots): Plot 1 determinism A vs determinism B; Plot2 recurrence A vs recurrence B (Orsucci, 2002)

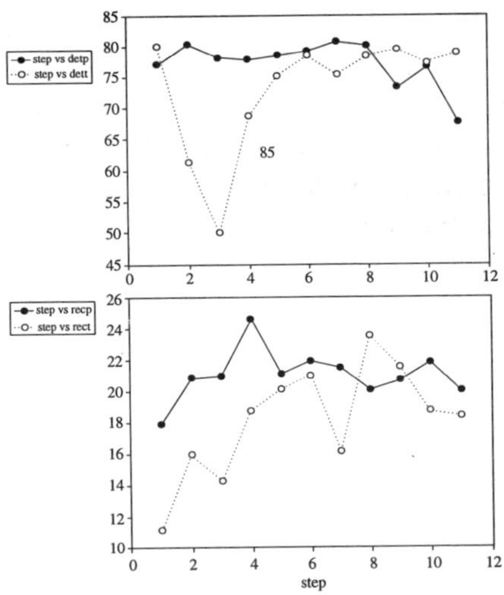

Controlled dialogue, Sample 2. A dialogue between a psychotherapist (black spots) and her therapeutic partner (white spots): Plot 1 detA vs detB; plot2 rec A vs rec B:

In the controlled-therapeutic dialogue we find at the beginning a distance between A and B, that tends to reduce towards the conclusion of the conversation. This final trend could be a slow start of synchronization.

Transcriptions have been coded and analyzed at the orthographic level using RQA: an informational conversation analysis that clarifies the "skeleton" of human interaction (Zbilut, 2002).

4.7 Cure environments

In ancient temples of Apollo and Dionysus, meditation retreats, shamanistic ceremonies there were rules and environmental conditions to protect mind-changing activities: incubation of dreaming, making prophecies, healing. In modern psychotherapies we have similar conditions to enhance the separation of the therapeutic system from outer variables and sources of noise, and foster the therapeutic coupling.

Even in the new experimental psychotherapeutic environments based on different forms of virtual reality we find similar conditions based on the physical requirements of synchronization and control theory. The setting of the psychotherapeutic office is designed for co-evolutionary change. The therapeutic system is partially insulated to protect it from outside interferences. It has a meeting schedule to maximize continuity and autonomy. It is centered on the therapeutic process and its needs.

The therapeutic environment should be considered as an artificial environment for the technology of mind change made by humans, for humans, with human resources. The most sophisticated and complex tool we still have to influence the complexities of mind functioning is still another mind: we should try to implement this situation at its best.

Being T mind and P mind isomorphic they start to co-evolve. Being both subsystems and the resulting therapeutic system nonlinear, they have just a partial capacity to drive the process. The T system has an intuition, based on previous experiences, about how his or her mind can interact with the Partner's kind of mind, and the possible results of this co-evolution.

Chapter 5: Co-Evolution

From Aither,
Time made a shining egg:
the progeny of Aither and Chaos.
Protogonos Theogony

5.1 Love Dynamics

We will try to continue on the path of the G-Draft, in trying to follow a dynamical model of human relations (Orsucci, 1998).

A way to approach the modelling of dynamics at work in deep relations is offered by the studies on the love relationship between the Renaissance poet Petrarch and his lady, Laura. This Platonic relation was source of inspiration for many poems but became also the origin of studies on the variations of their love and their mood, influenced by other variables (…mostly her husband). We will follow the works of Sergio Rinaldi (1996; 1996a; 1996b) on this historical paradigm.

FIG. 2. *Portraits of Laura and Petrarch (Biblioteca Medicea Laurenziana, ms. Plut. cc. VIIIv–IX, Florence, Italy; courtesy of Ministero per i Beni Culturali e Ambientali).*

Laura and Petrarch (Rinaldi, 1996b)

Romantic relationships are somehow the simplest case since they involve only two individuals. The most common approach to the problem is rooted in attachment theory (Bowlby, 1969; 1973; 1980) which explains why infants become emotionally attached to their primary caregivers and why they often experience emotional distress when separated from them. Empirical research has focused on different attachment styles (secure, anxious, avoidant...) in children (Ainsworth, 1978) and adult individuals. Several hypotheses have been generated about the nature and emotional quality of romantic relationships developed by people who exhibit different attachment styles.

Attachment styles can be regarded, in biophysical terms, as the different approaches of each human system to coupling and subsequent phase transitions implied in synchronization and co-evolution.

In conclusion, one can reasonably argue that the main features of love-stories should be largely dictated by the attachment styles of the individuals involved. Love-stories are dynamic processes that start from zero (two persons are completely indifferent one to each other when they first meet), develop (more or less quickly) and end up into some sort of regime. Real-life observations tell us that most of the times *transients* develop very regularly and *asymptotic regimes are stationary and associated to positive romantic relationships*. But there are also love-stories initially characterized by *stormy patterns of the feelings as well as by cyclic regimes*, like that identified by Jones (1995) in Petrarch's *Canzoniere,* his celebrated book of love poems. This reminds very much the behavior of dynamical systems tending toward equilibriums or limit cycles.

Observations also point out the *existence of multiple attractors*. For example, it is known that steady and high quality romantic relationships can turn into a state of permanent antagonism after a disturbance, for example after a temporary infatuation of one of the two partners for another person. Finally, even bifurcations can be naturally invoked if one looks at the effects of age and other social variables (power, money etc.), capable of transforming tempestuous relationships into steady ones or vice-versa.

5.1.1 ODEs

The above remarks suggest the *use of Ordinary Differential Equations* for modelling the dynamics of the feelings between two individuals. It is therefore not surprising if a few contributions have recently appeared along this line. They are rooted in a one page pioneering paper by Strogatz (1988) entitled "love affairs and differential equations".

The models proposed up to now are *minimal models*, in the sense that they have the lowest possible number of state variables, namely one for each member of the couple.

Such variables, indicated by x_1 and x_2, are a measure of the love of individual 1 and 2 for the partner. Positive values of x represent positive feelings, ranging from friendship to passion, while negative values are associated with antagonism and disdain. Complete indifference is identified by $x = 0$.

Minimal models are a crude simplification of reality. Because love is a complex mixture of different feelings (esteem, friendship, sexual satisfaction...) and can be hardly captured by a single variable. Because all the tensions and emotions involved in the social life of a person cannot be included in a minimal model. In other words, only the interactions between the two individuals are taken into account, while the rest of the world is kept frozen and does not participate explicitly in the formation of love dynamics. This means that rather than attempting to be complete, the aim is to check which part of the actions observed in real life can in theory be explained by the few ingredients included in the model.

5.1.2 *Oblivion, Return and Instinct*

Three basic processes, namely *Oblivion, Return* and *Instinct*, are assumed to be responsible of love dynamics. More precisely, the instantaneous rate of change, of individual's ***i*** love is assumed to be composed of three terms, i.e.,

$$\dot{X}_i = O_i + R_i + I_i$$

where the functions O_i, R_i and I_i, have different specifications in each model.

In the first model discussed in literature (Strogatz, 1988; Radzicki, 1993; Strogatz, 1994) oblivion and instinct are neglected, while return R_i, which interprets the reaction of individual i to the partner's love x_j, $j \neq i$, is assumed to be proportional to x_j. But the two proportionality coefficients have opposite sign, *i.e.,* one of the two lovers is a bit masochist and hates to be loved and loves to be hated. Thus, the model turns out to be a linear oscillator. Strogatz himself explains why he has made these very extreme assumptions: his goal was to teach harmonic oscillators using "a topic that is already on the minds of many college students: the time evolution of a love-affair".

In the two other models oblivion and instinct are also present. Oblivion is specified as:

$$O_i = -\alpha_i x_i$$

so that in the extreme case of an individual i who has lost the partner $(R_i = I_i = 0_i)$, x_i vanishes exponentially at a rate α_i. For this reason, α_i is called *forgetting coefficient.* Instinct I_i describes the reaction of individual i to the partner's appeal A_j. Of course, it must be understood that appeal is not mere physical attractiveness, but more properly and in accordance with evolutionary theory, a suitable combination of different attributes among which age, education, earning potential and social position. Moreover, there might be gender differences in the relative weights of the combination (Feingold, 1990; Sprecher et al., 1994).

In the second model (Rinaldi, 1996a) all processes are assumed to be linear and given by

$$O_i = -\alpha_i x_i$$
$$R_i = \beta_i x_j$$
$$I_i = \gamma_i A_j$$

with α_i, β_i and γ_i, $i = 1, 2$, positive. Thus, the model turns out to be a *positive linear system* (Luenberger, 1979; Berman *et al.,* 1989; Rinaldi and Farina, 1995) enjoying, as such, a number of remarkable properties. In particular, if the geometric mean reactivity to love ($\sqrt{\beta_1\beta_2}$) is smaller than the geometric mean forgetting coefficient ($\sqrt{\alpha_1\alpha_2}$) there is a unique positive and stable equilibrium

$$E^+ = (x_1^+ x_2^+)$$

and the two persons, completely indifferent one to each other when they first meet, develop a love story characterized by increasing feelings. Moreover, the quality (x_1^+ and x_2^+) of the romantic relationship at equilibrium improves with the reactivity to love (β_i) and appeal (γ_i). Finally, an increase of the appeal A_i of individual ***i*** gives rise to an increase of the feelings (x_1^+ ,x_2^+) of both individuals at equilibrium, but the relative improvement is higher for the partner of individual i (in other words, there is a touch of altruism in a woman [man] who tries to improve her [his] appeal).

Some of these individual properties can be used to infer properties at the community level: the main result along this line is that a community composed of N linear couples is stable if and only if the partner of the *n-th* most attractive woman of the community is the *n-th* most attractive man *(*n =1, 2, ...*N)*. This means that individual appeal is the driving force that creates order in our societies.

5.1.3 The third model

In the third model (Rinaldi and Gragnani, 1996) return $\boldsymbol{R}_i$ and instinct $\boldsymbol{I}_i$ still assumed to depend only upon x_j and α_j respectively, but the dependence is nonlinear: it takes into account the traits of so-called *secure individuals*. Secure individuals have positive mental models of themselves and of the others and their romantic relationships are characterized by intimacy, closeness, mutual respect and involvement. They react positively to the partner's love and are not afraid about someone becoming emotionally close to them. In conclusion, their return $\boldsymbol{R}_i$ is an increasing function of x_i.

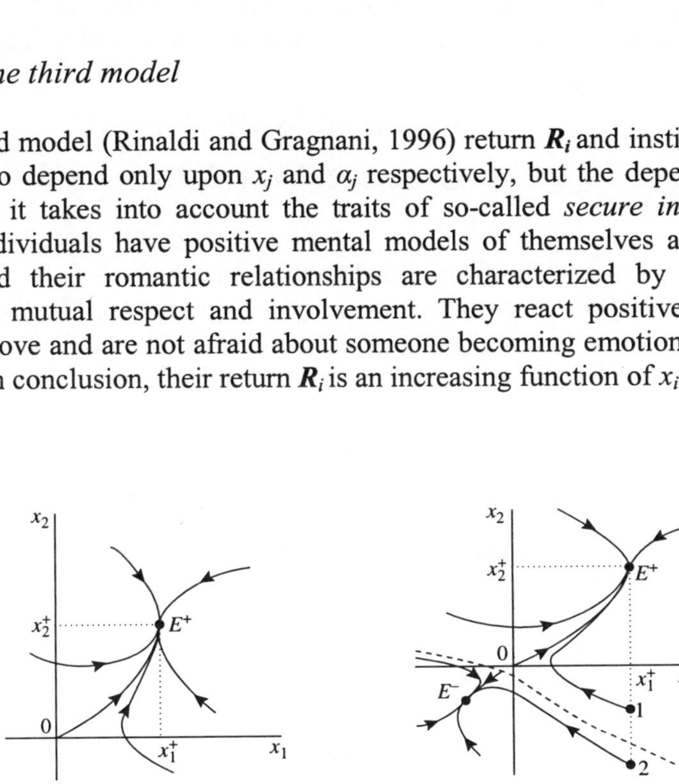

(a) (b)

Return functions of secure individuals (a) and non-secure individuals (b). (Rinaldi, 1996).

shows the graph of a typical return function of a secure individual: the same graph characterizes the function $I_1(A)$. Note that R^+_i and R^-_i denote the return for very large positive and negative partner's feelings. Boundedness is a property that holds also for non-secure individuals: it interprets the psycho-physical mechanisms that prevent people from reaching dangerously high stresses. By contrast, the fact that the function increases with x_j is typical of secure individuals, since non-secure individuals react negatively to too high pressures and involvement (Griffin and Bartholomew, 1994). The model is therefore

$$x_1 = -\alpha_1 x_1 + R_1^*(x_2) + I_1^*(A_2)$$

$$x_2 = -\alpha_2 x_2 + R_2^*(x_1) + I_2^*(A_1)$$

where the graphs of the functions R^* and I^* are like in Fig. la. This model retains many of the properties of the simpler linear model, like the existence of a stable positive equilibrium E^+, and allows one to derive the same conclusions on the role of the appeals at community level.

5.1.4 Romantic and fragile

The main difference is that the *nonlinearities can give rise to a negative stable equilibrium* E^- (generated *through a fold bifurcation*). Couples with a unique attractor (E^+) are called *robust,* while couples with two attractors (E^+ and E^-) are called *fragile.*

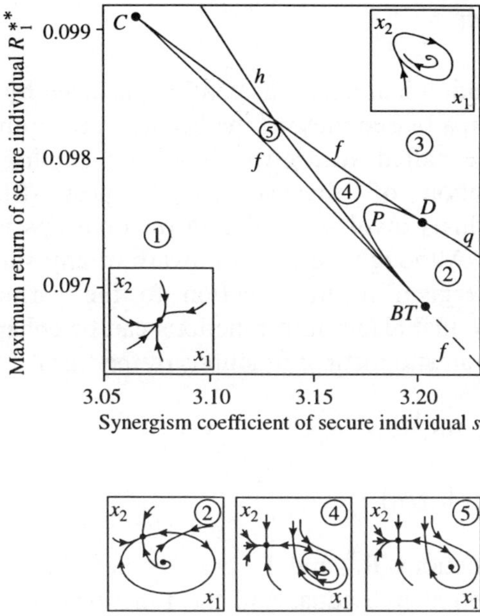

State portraits of couples composed of secure individuals: robust couples (a) and fragile couples (b). In (b) the boundary of the two basins of attraction (dashed line) is the stable manifold of a saddle (Rinaldi, 1996).

This figure shows the two corresponding state portraits. Note that fragile couples can switch from E^+ to E^- if one of the two individuals has a sudden drop in interest for the partner (see trajectories starting from points 1 and 2 in Fig. B).

We stress that limit cycles cannot exist in model even if individuals are non-secure, i.e., even if the functions R* are like in Fig. lb. In fact, the divergence of the system, equal to - $(\alpha_1 + \alpha_2)$, does not change sign and Bendixon's criterion implies the non-existence of limit cycles. This means that at least one individual ***i*** of the couple with cyclical love dynamics must have an instinct function I_i depending also upon x_1 or a return function R_i depending upon both state variables. This is related with synergism.

5.1.5 Synergism

It is known that individual reactions can be enhanced by love. For example, mothers have often a biased view of the beauty of their children. This kind of phenomenon, here called *synergism,* has been empirically observed in a study on perception of physical attractiveness (Simpson, 1990) by comparing individuals involved in dating relationships with individuals not involved in them. Although we are not aware of any study pointing out the existence of synergism in the reaction to the partner's love, we can reasonably assume that also return functions can be enhanced by love. Thus, we consider reaction and instinct functions of the form

$$R_i = (1 + S_i^R(x_i))R_i^*(x_j)$$

$$I_i = (1 + S_i^I(x_i))\, I_i^*(A_j)$$

where the functions R_i^* and I_i^* are, by definition, the reactions of a completely indifferent individual and the functions $S_i^R(x_i)$ and $S_i^I(x_i)$ are zero for to $x_i \leq 0$ and increasing, convex/concave and bounded for $x_i > 0$. The upper bounds of the functions S_i^R and S_i^I are indicated by s_i^R and s_i^I and are called synergism coefficients.

We have shown in the preceding section that couples composed of non-synergic individuals, *i.e.,* couples with $s_i^R = s_i^I = 0$, cannot have cyclic behavior. Thus, synergism is necessary for generating complex love dynamics. It can be shown however, that synergism is not sufficient if the couple is composed of secure individuals *(i.e.,* individuals characterized by reaction and instinct functions R^* and I^* like in Fig. A). We prove this result by referring to the case of synergic instinct functions, *i.e.,* by analyzing the model

$$x_1 = -\alpha_1 x_1 + R_1^*(x_2) + (1 + S_1^I(x_1))\, I_1^*(A_2)$$

$$x_2 = -\alpha_2 x_2 + R_2^*(x_1) + (1 + S_2^I(x_2))\, I_2^*(A_1)$$

5.2 Dynamical love

We have seen how dynamics of love between two persons can be investigated by means of a minimal model composed of two differential equations, one for each individual. Three mechanisms of love growth and decay have been taken into account: the forgetting process, the pleasure of being loved and love itself (the reaction to partner's appeal). This has been done by introducing two functions, called return and instinct functions, which differ in the cases of secure and non-secure individuals. The fact, here called synergism, which a woman [man] might react more strongly when she [he] is in love, has also been modeled. As a result, individuals are secure or non-secure, and synergic or non-synergic, for a total of four different classes. Thus, it has been shown that couples composed of non-synergic individuals as well as couples composed of secure individuals cannot have complex dynamics. By contrast, couples composed of a secure and synergic individual and a non-secure and non-synergic individual can have complex dynamics. In other words, *the coexistence of synergism and insecurity within the couple is the reason for tempestuous romantic relationships. This result has been proved through the analysis of the local and global bifurcations of the model. Fold, Hopf and homoclinic bifurcations have been detected, in accordance with the existence of a Bogdanov-Takens codimension-2 bifurcation point.*

The bifurcation diagrams can be easily interpreted in terms of individual behavior and appeal. They show that, with the exception of a narrow band in parameter space (regions 4 of Fig B), the system has a unique global attractor: equilibrium or a limit cycle. Both attractors are associated to positive feelings, *i.e.,* to a satisfactory quality of the romantic relationship. In general, the attractor is equilibrium when the individuals are not too sensitive, *i.e.,* when their synergism and reactivity to partner's love and appeal are, as a whole, low. By contrast, *high sensitivity (as well as high attractiveness) implies cyclic behavior, as one would intuitively expect.*

Our bifurcation diagrams also show that *aging has a stabilizing effect.* Indeed, it is generally believed that individual appeal, as well as synergism and reactions to partner's love and appeal, slowly deteriorate with aging. Thus, couples with complex love dynamics (region 3) can slowly vary during their life and finally become stationary by crossing the bifurcation curves h, p or q. The transformation is accompanied by a reduction of the amplitudes of the emotional ups and downs of a cyclic regime (when Hopf bifurcation curve h is approached) or by a reduction of the frequency of these ups and downs (when homoclinic curves p and q are approached).

As for any minimal model, the extensions one could propose are innumerable. In accordance with the most recent developments of attachment theory, learning and adaptation processes could be taken into account allowing for some behavioral parameters to slowly vary in time. Suitable nonlinearities could be introduced in order to develop theories for particular subclasses of non-secure individuals. Men and women could be distinguished by using different state equations. The dimension of the model could also be enlarged in order to consider individuals with more complex personalities or the dynamics of love in larger groups of individuals (e.g. families, communities etc.).

5.3 Synchronization

As seen in the previous chapter, we can study these dynamical behaviors in terms of coupling and synchronization.

Synchronization of coupled oscillating systems means the appearance of certain relations between their phases and frequencies. It can be used to modify a system by controlled interaction, and also, in the reverse sense to detect some features of the coupling structure. Synchronization can happen between periodic, noisy or chaotic systems. Recently, with increase of studies of chaotic oscillations, the notion of synchronization has been generalized to the latter case. In this context, different phenomena exist which are usually referred to as *synchronization*, so one needs a more precise description to specify them.

Due to a strong interaction of two (or a large number) of identical chaotic systems, their states can coincide, while the dynamics in time remains chaotic. This case can be denoted as *complete synchronization* of chaotic oscillators. It can be easily generalized to the case of slightly non-identical systems, or the interacting subsystems. A different approach is based on the calculation of the attractor dimension of the whole system and its comparison with the partial dimensions calculated in the phase subspaces formed by the coordinates of each interacting oscillator.

Another well-studied effect is the *chaos destroying synchronization*, when periodic external force acting on a chaotic system destroys chaos and a periodic regime appears (or, in the case of an irregular forcing, the driven system follows the behavior of the force). This effect occurs for a relatively strong forcing as well. A characteristic feature of these phenomena is the existence of a *threshold coupling value* (depending on the Lyapunov exponents of individual systems). In another approach, an *overlapping of power spectra* of certain observables in the interacting systems has been studied. It has been shown that due to interaction, the widths of the peaks of the power spectra may become practically equal, and the peaks frequencies become closer. Because of an analogy to the coincidence of frequencies of synchronized periodic systems this effect has been interpreted as synchronization and has been quantified by means of cross-correlation functions and cross-spectra. The effect of phase synchronization of chaotic systems is mostly close to synchronization of periodic oscillations, where only the phase locking is important, while no restriction on the amplitudes is imposed.

Thus, *we define phase synchronization of chaotic system as the appearance of a certain relation between the phases of interacting systems (or between the phase of a system and that of an external force), while the amplitudes can remain chaotic and are, in general, non-correlated.*

Of course, the very notion of phase and amplitude of chaotic systems is rather non-trivial. Roughly speaking, the phase of an autonomous self-sustained oscillatory system is related to the symmetry with respect to time shifts. Therefore, the phase disturbances do not grow or decay, what corresponds to the zero Lyapunov exponents. If the oscillations are periodic, the phase rotates nearly uniformly, while in the chaotic case the dynamics of the phase is affected by chaotic changes of the amplitude, so one can expect a Brownian (random-walk-like) behavior of the phase.

5.3.1 Neural synchronization

Synchronization is related to several central issues in neuroscience, for neural *bindings* within the same area or for communication between different areas of the CNS. Brain activity is observed at three main space-time scales.

- At the *microscopic scale* are the pulses (action potentials) of single neurons measured in milliseconds and microns.

- At the *macroscopic scale* are the domains of high metabolic demand that are imaged in seconds and centimeters (Freeman, 2000; 2002) by a variety of techniques for measuring the spatial patterns of cerebral blood flow.

- In between, at the *mesoscopic scale* of millimeters (mm) and tenths of a second are the patterns of the massed dendritic potentials seen in electroencephalographic (EEG) recordings from waking and sleeping brains.

These waves have oscillations in several low frequency ranges, such as theta (3-7 Hz), alpha (8-12 Hz), beta (13-30 Hz), and gamma (30-100 Hz). Long stretches of recordings give broad spectra roughly conforming to $1/f^2$, which have suggested to many researchers the possibility that this activity is governed by chaotic attractors instead of limit cycles embedded in noise. This interesting hypothesis has been pursued with the aid of models derived from deterministic chaotic systems, such as twist-flip maps, Lorenz, Rössler and Chua attractors. These low-dimensional models have engendered highly successful algorithms such as those of Takens, Grassberger, Procaccia and Guckenheimer for estimating the correlation dimensions and Lyapunov exponents of these models. This success has led to efforts to model the cognitive functions of sensory cortices in the realm of pattern classification with networks of chaotic elements that are coupled by nearest neighbor or next to nearest neighbor with weighted connections. The proper response of such systems would be synchronization or phase locking of an array of chaotic elements into one of a set of global chaotic attractors, after suitable training to form the attractor landscape.

Attempts to develop and apply such models by calculating correlation dimensions and related measures from EEGs have not succeeded in converging to reliable. The failure is partly due to an assumption that the stationarity and autonomy of the models holds also for brain systems, which is inappropriate because *brains are continuously engaged with their bodies and environments, and they undergo repeated state transitions in multiple time scales*. Most importantly, the methods fail because the EEG traces recorded from brains are heavily invested with *various kinds of noise*, for some of which the dimensions are infinite. Estimates of correlation dimension invariably increase with the duration of the sample.

EEGs, anyway, manifests a small fraction of the total variance of the neurons in the generating populations that is covariant (synchronous), and the spatial pattern of phase reveals a critical determinant of the size of the areas of self-organized activity in sensory cortices (~ 1 cm). The group velocity of the state transitions by which spatiotemporal patterns of cortical activity emerge can be successfully extracted from noise by the targets of cortical transmissions (Freeman, 2000).

5.4 Synchronization and control

Pecora and Carroll suggested that it is possible *to plan the synchronization of chaotic system by introducing appropriate coupling between them: in this way controlling them*. They have shown that if a reproduced part (Response or subsystem) of chaotic system responses to a chaotic signal from the full (Drive) system, under some conditions the signal in the response part would converge to the corresponding signals in the drive system. Here, in the following scheme, the drive or master system controls the response or slave system through the replica part (Pecora, 1990; Pecora, 19 91; Carroll, 1993; Abarbanel, 1994).

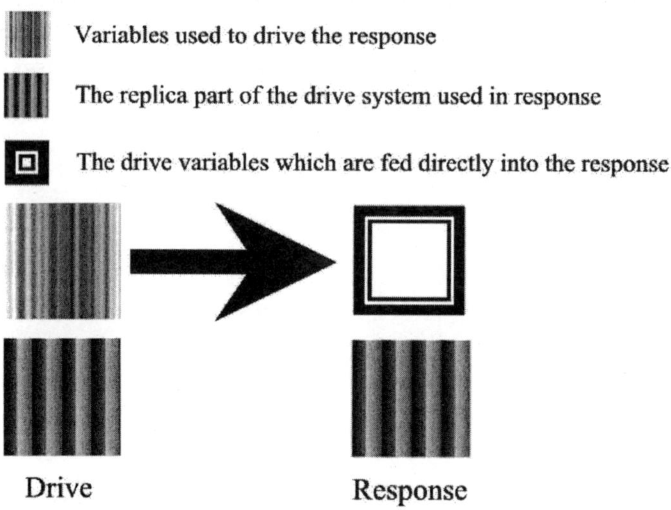

Schematic of drive-response scenarios of chaos synchronization

This is called synchronization of chaotic systems, and further this idea has been generalized by *cascading* the reproduced parts or subsystems. Here the considered chaotic system may be divided into two subsystems, each synchronizing with the full system, when driven by the appropriate chaotic signal. By arranging one of these synchronized subsystems to drive the other subsystem, it is possible to produce synchronized signals in the subsystem.

This cascading can eventually produce feedback loops in the master system, as usually happens in living systems, including humans. This generates a *co-evolution* of chaotic systems that can be driven by an appropriate *selection in the natural rich variation of states in chaotic systems*. That is a matter of *controlling the co-evolution.*

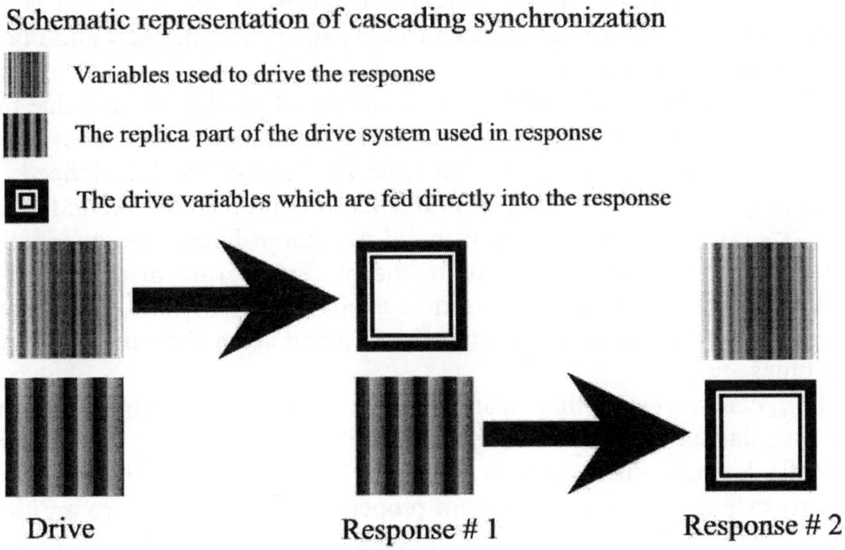

5.4.1 *Controlling chaos*

The control of chaos: exploiting the critical sensitivity to initial conditions to play with chaotic systems A deterministic system is said to be chaotic whenever its evolution sensitively depends on the initial conditions. This property implies that two trajectories emerging from two different closely initial conditions separate exponentially in the course of time. The necessary requirements for a deterministic system to be chaotic are that the system must be nonlinear, and be at least three dimensional.

The fact that some dynamical model systems showing the above necessary conditions possess such a critical dependence on the initial conditions was known since the end of the last century. However, only in the last thirty years, experimental observations have pointed out that, in fact, chaotic systems are common in nature. They can be found, for example, in Chemistry (Belouzov-Zhabotinski reaction), in Nonlinear Optics (lasers), in Electronics (Chua-Matsumoto circuit), in Fluid Dynamics (Rayleigh-Bénard convection), etc. Many natural phenomena can also be characterized as being chaotic. They can be found in meteorology, solar system, heart and brain of living organisms and so on.

Due to their critical dependence on the initial conditions, and due to the fact that, in general, experimental initial conditions are never known perfectly, these systems are intrinsically unpredictable. Indeed, the prediction trajectory emerging from a *bonafide* initial condition and the real trajectory emerging from the real initial condition diverge exponentially in course of time, so that the error in the prediction (the distance between prediction and real trajectories) grows exponentially in time, until making the system's real trajectory completely different from the predicted one at long times.

For many years, this feature made chaos undesirable, and most experimentalists considered such characteristic as something to be strongly avoided. Besides their critical sensitivity to initial conditions, chaotic systems exhibit two other important properties. Firstly, there are an infinite number of unstable periodic orbits embedded in the underlying chaotic set.

In other words, the skeleton of a chaotic attractor is a collection of an infinite number of periodic orbits, each one being unstable. Secondly, the dynamics in the chaotic attractor is ergodic, which implies that during its temporal evolution the system ergodically visits small neighborhood of every point in each one of the unstable periodic orbits embedded within the chaotic attractor.

A relevant consequence of these properties is that a chaotic dynamics can be seen as shadowing some periodic behavior at a given time, and erratically jumping from one to another periodic orbit. The idea of controlling chaos is then when a trajectory approaches ergodically a desired periodic orbit embedded in the attractor, one applies small perturbations to stabilize such an orbit. If one switches on the stabilizing perturbations, the trajectory moves to the neighborhood of the desired periodic orbit that can now be stabilized. This fact has suggested the idea that the critical sensitivity of a chaotic system to changes (perturbations) in its initial conditions may be, in fact, very desirable in practical experimental situations. Indeed, if it is true that a small perturbation can give rise to a very large response in the course of time, it is also true that a judicious choice of such a perturbation can direct the trajectory to wherever one wants in the attractor, and to produce a series of desired dynamical states. This is exactly the idea of targeting.

The important point here is: *because of chaos, one is able to produce an infinite number of desired dynamical behaviors (either periodic or not periodic) using the same chaotic system, with the only help of tiny perturbations chosen properly. We stress that this is not the case for a non-chaotic dynamics, wherein the perturbations to be done for producing a desired behavior must, in general, be of the same order of magnitude as the unperturbed evolution of the dynamical variables.*

The idea of chaos control was enunciated at the beginning of this decade at the University of Maryland. The ideas for controlling chaos were outlined and a method for stabilizing an unstable periodic orbit was suggested, as a proof of principle. The main idea consisted in waiting for a natural passage of the chaotic orbit close to the desired periodic behavior, and then applying a small judiciously chosen perturbation, in order to stabilize such periodic dynamics (which would be, in fact, unstable for the unperturbed system). Through this mechanism, one can use a given laboratory system for producing an infinite number of different periodic behavior (the infinite number of its unstable periodic orbits), with a great flexibility in switching from one to another behavior. Much more, by constructing appropriate goal dynamics, compatible with the chaotic attractor, an operator may apply small perturbations to produce any kind of desired dynamics, even not periodic, with practical application in the coding process of signals. The perspective of unifying the techniques of deterministic chaos control with a statistical stochastic description as a possible therapeutic strategy against dynamical diseases is surely something to consider.

5.5 Controlling co-evolution

John von Neumann, circa 1950, affirmed: “All stable processes, we shall predict. All unstable processes, we shall control.” (Dubé, JL & Desprès, P, 2000)

In his 1985 Giord Lectures, Freeman Dyson expressed his quite different opinion on the matter of chaos:

“A chaotic motion is generally neither predictable nor controllable. It is unpredictable because a small disturbance will produce exponentially growing perturbation of the motion. It is uncontrollable because small disturbances lead only to other chaotic motions and not to any stable and predictable alternative. Von Neumann's mistake was to imagine that every unstable motion could be nudged into a stable motion by small pushes and pulls applied at the right places.

It was not until 1990 however that Ott, Grebogi and Yorke (OGY) addressed the question of control of chaos and described, very much in the spirit of von Neumann, the theoretical steps necessary to achieve this goal (Boccaletti, 2001; 2000; Pikovsky, 2001). This work was rapidly followed by experimental verification: von Neumann's dream had becoming reality?

It is reasonable to assume that one does not have complete knowledge about the system dynamics since our system is typically complicated and imperfect. It is better, then, to work in the space of solutions since the equations, even if available, are not too useful due to the sensitivity of the dynamics to perturbations. *One gets solutions by obtaining a time series of one dynamically relevant variable.* The *right perturbation*, therefore, to be applied to the system is selected *after a learning time*, wherein the dependence of the dynamics on some external control is tested experimentally. Such perturbation can affect either a control parameter of the system, or a state variable.

Let us draw the attention on a chaotic dynamics developing onto an attractor in a D-dimensional phase space. One can construct a section of the dynamics such that it is perpendicular to the chaotic flow (it is called Poincaré section). This (D-1)-dimensional section retains all the relevant information of the dynamics, which now is seen as a mapping from the present to the next intersection of the flow with the Poincaré' section. Any *periodic behavior* is seen here as a periodic cycling among a discrete number of points (the number of points determines the periodicity of the periodic orbit). Since all periodic orbits in the unperturbed dynamics are unstable, also the periodic cycling in the map will be unstable. Furthermore, since, by ergodicity, the chaotic flow visits closely all the unstable periodic orbits, this implies that also the mapping in the section will visit closely all possible cycles of points corresponding to a periodic behavior of the system. Let us then consider a given periodic cycle of the map, such as period one. A period one cycle corresponds to a single point in the Poincaré section, which repeats itself indefinitely. Now, *because of the instability of the corresponding orbit, this point in fact possesses a stable manifold and an unstable manifold.* For stable (unstable) manifold we mean the collection of directions in phase space through which the trajectory approaches (diverges away from) the point geometrically.

The control of chaos idea consists in perturbing a control parameter when the natural trajectory is in a small neighborhood of the desired point, such that the next intersection with the Poincaré section puts the trajectory on the stable manifold. In this case, all divergences are cured, and the successive natural evolution of the dynamics, except for nonlinearities and noise, converges to the desired point (that is, it stabilizes the desired periodic behavior). Selection of the perturbation is done by means of a reconstruction from experimental data of the local linear properties of the dynamics around the desired point.

In some practical situations, however, it may be desirable to perform perturbations on a state variable accessible to the operator.

This suggests the development of some alternative approaches. The first consists in designing a *proper feedback* line through which a state variable is directly perturbed such as to control a periodic orbit.

The second method requires the availability of a state variable for observation and for the perturbations. In such a case, a *negative feedback* line can be designed which is proportional to the difference between the actual value of the state variable, and the value delayed of a time lag T. The idea is that, when T coincides with the period of one unstable periodic orbit of the unperturbed system, the negative feedback pushes to zero the difference between the present and the delayed dynamics, and the periodic orbit is stabilized. Furthermore, as soon as the control becomes effective, this difference goes effectively to zero, so that the feedback perturbation vanishes. Moreover, as before, a *preliminary learning time* is needed, for learning the periods of the unstable periodic orbits. In the above mechanism, the *proportionality constant* entering in the feedback loop is given in where an adaptive technique has been introduced which automatically selects this constant by adaptively exploiting the local dynamics of the system (the Occasional Proportional Feedback OPF method and the Recursive Proportional Feedback RPF method).

5.5.1 OGY

The idea of Ott, Grebogi, and Yorke (OGY) is that *chaos may indeed be desirable* since it can be controlled by using small perturbation to some accessible parameter or to some dynamical variable of the system.

The major key ingredient for the control of chaos is the observation that a chaotic set, on which the trajectory of the chaotic process lives, is embedded within it a large number of unstable low-period periodic orbits. In addition, because of ergodicity, the trajectory visits or accesses the neighborhood of each one of these periodic orbits. Some of these periodic orbits may correspond to a desired system's performance according to some criterion. The second ingredient is the realization that chaos, while signifying sensitive dependence on small changes to the current state and henceforth rendering unpredictable the system state in the long time, also implies that the system's behavior can be altered by using small perturbations. Then, the accessibility of the chaotic systems to many different periodic orbits combined with its sensitivity to small perturbations allows for the control and the manipulation of the chaotic process. Specifically, the OGY approach is then as follows. One first determines some of the unstable low-period periodic orbits that are embedded in the chaotic set. One then examines the location and the stability of these orbits and chooses one which yields the desired system performance. Finally, one applies small control to stabilize this desired periodic orbit. However, all this can be done from data by using nonlinear time series analysis for the observation, understanding and control of the system. This is particularly important since chaotic systems are rather complicated and the detailed knowledge of the equations of the process is often unknown.

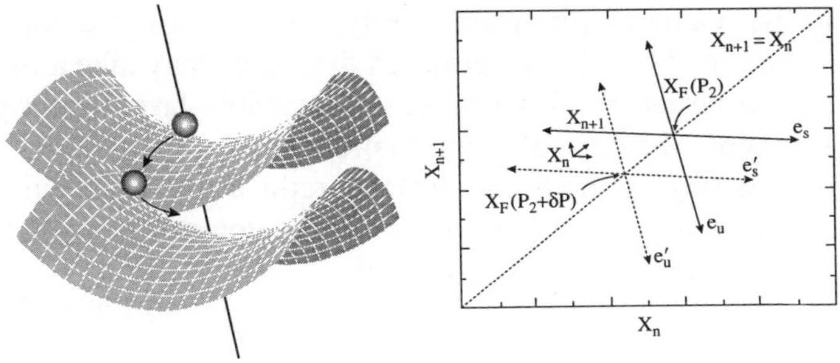

Local geometry of control: left *2D saddle dynamics* and right *linearization of the stable and unstable manifolds* (Dubè, 2000)

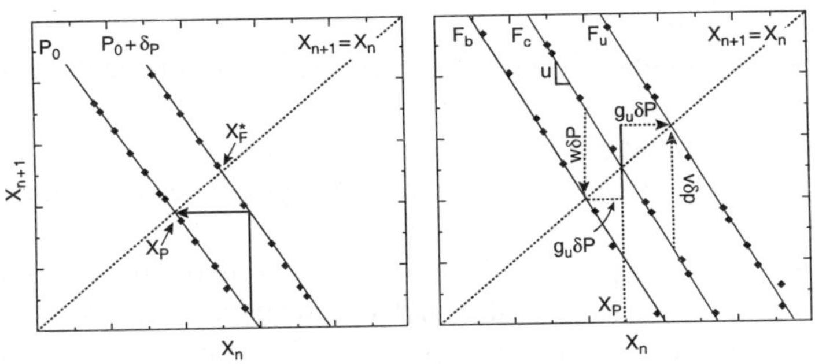

Alternative local geometries of control: left the Occasional Proportional Feedback (OPF method) and right the Recursive Proportional Feedback (RPF method) (Dubè, 2000)

5.6 Variation and selection

Targeting desirable states within chaotic attractors is a preliminary of selection and controlling co-evolution. One of the major problems in the above process is that one can switch on the control only when the system is sufficiently close to the desired behavior. This is warranted by the ergodicity of chaos regardless of the initial condition chosen for the chaotic evolution, but it may happen that the small neighborhood of a given attractor point (target) may be visited only infrequently, because of the locally small probability function. Thus the unperturbed dynamics may take a long time to approach a given target, resulting in an unacceptably large waiting time for the operator to apply the control of chaos process.

Efficient targeting methods can, instead, reduce the waiting time, and so they can be seen as a preliminary task for chaos control, independent of the particular control algorithm that one applies.

5.6.1 Learning from experience

The lessons learned through the previous examples and many more not reported here allow us to draw a list of the properties and advantages of the adopted philosophy on the co-evolution of complex systems and to point to remaining difficulties:

- no model dynamics is required a priori and only local information is needed;
- computations at each step are minimal;
- gentle touch: the required changes in can be quite small (<1%);
- multi-purpose flexibility: different periodic orbits can be stabilized for the same system in the same parameter range;
- control can be achieved even with imprecise measurements of eigenvalues and eigenvectors: the methods are robust;
- the methods can also be applied to synchronization of several different chaotic systems.

At least three complications come to mind when one considers the implementation of chaos control strategies:

- The presence of noise (dynamical or environmental) may induce loss of control or hinder it altogether.
- The average waiting time to fall in the neighbourhood may be very (too) long, especially for Hamiltonian systems and a targeting strategy should complement the control method.
- The system's parameters may drift with time and this non-stationarity should be accounted for by updating the control information. Tracking is the name given to this procedure.

- Cascading can generate feedbacks and lead to co-evolution, and this can generate uncontrollable drifts in the synchronized systems.
- The drive system has to re-split in an observing-controlling subsystem each time a feedback cascading happens.

However, the greatest challenge will remain for some time the application to complex biological systems: in particular to mind and brain dynamics. Despite early efforts, euphoria has been replaced by pragmatism.

The perspective of unifying the techniques of deterministic chaos control with a statistical description as a possible therapeutic strategy against dynamical diseases is the challenge for next years.

Chapter 6: Change

Time is not a line
but a network of intentionality.
Maurice Merleau-Ponty

6.1 Transitions

Coupling can start with intermittencies and then become stable and constant. Synchronization is usually preceded by a dynamical clustering, that can be glassy. The sequence of transitions can usually be:

Turbulence > intermittency > glassy states> coherency

The addition of noise to the global system can foster the synchronization process (mostly between weak attractors), but beyond a certain threshold it can induce uncontrolled turbulences. Coupling, in any case, increases the number of degrees of freedom of the resulting system. This, besides any specific control, is very useful to cure any dynamical disease on the system.

A fractal intermittency has been observed on the threshold of weak synchronizations, while strong synchronizations generate smoother landscapes (Pikovsky, 2001).

6.1.1 Symbiosis

Symbiosis is another mechanism of meta-system transitions, and an equivalent of coherency in coupling. Non-similar systems may start cooperating, and this cooperation represents semiotic closure at a higher level. For example, lichens are symbiotic organisms that originated from fungi and algae; eukaryotic cells originated from a symbiosis of several types of prokaryotic cells.

When two autocatalytic systems cooperate, they produce resources for each other. *Resources are streams of signs: thus, cooperation is a semiotic relationship. Each component has a double role in such a system. First, it is self-reproducing on its own (self-interpretation), and second, it produces signs that are interpreted by another component. The major obstacle on the way of cooperation is a possible evolutionary instability* (Hoffmeyer, 1997; Margulis, 1981; 1998).

Let us consider cooperating chemical species A and B that produce resources for each other. Species A may mutate into a selfish species A_1 which will use resources produced by species B without providing help to the species B. As a result, the cooperation between species is broken.

Cooperation is evolutionary stable only if specific restrictions are applied on communication (resource exchange). We could call these restrictions *encapsulation*. For example, several representatives of species A and B may form small groups, so that communication occurs only among members of a group. Hierarchical systems have several levels of *encapsulation* and this makes them more stable than systems that have no communication restrictions. Encapsulation can be considered as a *defence mechanism* in systemic terms.

The relationships between populations of two interacting species, A and B, can be described by making a chart of all possible types of interactions. Interactions may be positive (+), (a species is positively affected by the interaction) or negative (-), (a species is negatively affected by the interaction). All the possible interactions can be described by the following chart:

	A	B	
Neutralism	0	0	Populations of A and B are independent and don't interact.
Mutualism	+	+	Both species benefit from the relationship.
Commensalism	+	0	One species benefits from the relationship but the other is neither harmed nor benefited
Competition	-	-	Both species have a negative effect on each other since they are competing for the same resources.
Predation	+	-	The predator (A) benefits while the prey (B) is negatively affected.
Parasitism	+	-	The parasite (A) benefits while the host (B) is negatively affected.

Commensalism and mutualism are often grouped together and called symbiosis.

Many interesting co-evolutionary relationships are mutualisms. Unlike competition, predation, or parasitism (notice that predation and parasitism share the same markers), mutualistic relationships tend to be quite unstable. Slight perturbations in conditions or population numbers and one species or the other could become extinct. It is thought that the stability of the environment permits mutualisms to be stable.

We should not invoke co-evolution without reasonable evidence that the traits in each species were a result of or evolved from the interaction between the two species. There are a variety of different modes of co-evolution. In some cases co-evolution is quite specific such as those between two cellular functions. The *endo-symbiont theory* proposes that current day mitochondria and chloroplasts were once free-living unicellular individuals. These cells entered the cytoplasm of other cells, an example of the general phenomenon of endosymbiosis. Current-day mitochondrial and chloroplast genomes are much smaller than the genome sizes of their presumed free-living ancestors. Some of this reduction in genome size is due to the transfer of genes from organelle genomes to the nuclear genome. Thus, being in the cellular environment has influenced the evolution of organelle genomes. There is evidence that the faster rate of evolution of animal mitochondrial DNA has accelerated the rate of evolution of some of the nuclear genes that function in the mitochondria.

6.1.2 Fitness landscapes

Kauffman (Kauffman, 1993; and in Palombo, 1999) suggests that no general law is likely to govern all possible non-equilibrium systems; the central aim is to define the general conditions that allow evolution and co-evolution to assemble complex systems. Not all complex systems can be assembled by an evolutionary process. The general character of *fitness landscapes* of co-evolutionary processes in which the adaptive moves of one *agent* deforms the fitness landscapes of its partners, and tries to show that an endogenous co-evolutionary process allows *agents*, each adapting for its own *selfish fitness*, to tune the couplings and ruggedness of their fitness landscapes such that the entire system achieves a specific self-organized critical state. Here, as if by an invisible hand, the fitness of each agent appears to be maximized, its probability of extinction appears to be minimized, yet a power law distribution of small and large avalanches of extinction and speciation events propagates through the model ecosystem.

6.2 The Adjacent Possible

The *Adjacent Possible is the route to change.* If we consider the possible states of a given coupling system, we can also consider those set of states that do not exist, but that are one step away from states that currently do exist. The set of kinds of states that are one step away from those that do exist constitutes the *adjacent possible.*

Since the work of John von Neumann, "games" have been a scientific metaphor for a much wider range of human interactions in which the outcomes depend on the interactive strategies of two or more persons, who have opposed or at best mixed motives. Among the issues discussed in game theory are:

1) What does it mean to choose strategies rationally when outcomes depend on the strategies chosen by others and when information is incomplete?

2) In "games" that allow mutual gain (or mutual loss) is it "rational" to cooperate to realize the mutual gain (or avoid the mutual loss) or is it "rational" to act aggressively in seeking individual gain regardless of mutual gain or loss?

3) If the answers to 2) are "sometimes", in what circumstances is aggression rational and in what circumstances is cooperation rational?

4) In particular, do ongoing relationships differ from one-off encounters in this connection?

5) Can rules of cooperation emerge spontaneously from the interactions of rational egoists?

6) How does real human behavior correspond to "rational" behavior in these cases?

7) If it differs, in what direction? Are people more cooperative than would be "rational?"

Recent developments in game theory, especially the award of the Nobel Memorial Prize in 1994 to three game theorists have revived the interest on this domain. One of the three was the inventor of the *Nash Equilibrium*, the mathematician who inspired novels and movies. If there is a set of strategies with the property that no player can benefit by changing her strategy while the other players keep their strategies unchanged, then that set of strategies and the corresponding payoffs constitute the Nash Equilibrium. The Nash equilibrium is a low-price, zero-profit equilibrium. But in some ways, Tucker's invention of the Prisoners' Dilemma example was even more important. Both are very useful to understand aspects of the (therapeutic) partnership (Goldstein,1994; Guastello, 1995; 2002).

6.2.1 Alliance for cure

In reviewing about fifty years of research in psychotherapy Horst Kachele (2001) concluded that the Therapeutic Alliance, the analog of *cooperation* in the analytic psychotherapy domain has become the most explored clinical concept. It is now evident that the construct of this alliance has both a direct and an indirect impact on outcome, i.e. the desired modifications in the partner system. There is agreement that the therapeutic alliance is a multi-dimensional construct composed of four relatively independent dimensions (11):

a) the partner's capacity to purposefully work in therapy
b) the partner's affective bond to the therapist
c) the therapist's empathic understanding and involvement
d) the partner and the therapist's agreement on treatment goals and tasks.

Kachele stresses that, *beyond any technical consideration "match between analyst and patient is a key predictor of outcome"*.

That is very important because, being all the bulk of possible variables beyond any illusion of total control, in clinical applications we should start from the most favorable condition for a possible cooperation. This can be defined: mastering random evolution by selection

It implies that every therapist knows, before starting a new treatment:

1. the kind of psychopathology he works better with;
2. the kind of personality we works better with;

a. if he or she "likes" the possible partner (sympathy and empathy).

This could be seen as something of non scientific or pre-scientific, but we assume that a sympathy and emotional resonance between the candidates to joint explorations in the co-evolutionary landscapes, are the markers of the possible empathy (collaboration) to be developed, in both directions. In other terms, the T system and the P system have some kind of resonance that precedes and sets conditions for any possible co-evolution.

This kind of precondition for a good collaboration can be generalized to natural couplings in everyday life: clinical contexts represent just a partially artificial subset.

Hans Sachs used to say that "even the most successful treatment is nothing more than an indent on a continent's surface". Remembering the role of sensitive dependence on initial conditions, the task is to put the "indent" at the right moment in the right place, but this has to be done after respecting all the targeting preconditions (waiting time etc.). The OGY method (and correlates) of gentle right timed pushes is the key.

According to Fonagy (1999), "it is important for the analyst to be aware that the changes sought are not changes in the patient's awareness of past events but rather changes in procedural and implicit memory. Thus recovery of past experience may be helpful but the understanding of current ways of being with the other is the key to change. For this, both self and other representation may need to alter and this can only be done effectively in the here and now." The core of change is in the therapeutic relation, because the therapeutic partner is gradually forced to consider it as a gymnasium specialized for life. The new coupling becomes a new system in itself and it reflects just partially the history of the separated subsystems. The work on synchronization is just on a very brief time window.

The transference and counter-transference represent what in the technical language of psychotherapy we have until now called *coupling* in human relations. We should consider the psychotherapeutic situation as a specialized setting in which a coupling between human subjects evolves under artificial and controlled conditions.

This protected environment allows a partial reduction of the degrees of freedom in the therapeutic system, but the way this coupling evolves has anyway multidimensional determinants. The study of the informational layers has many advantages:

1. Is susceptible of statistico-mathematical treatments;
2. Is beyond the conscious and unconscious control of the subjects;
3. In heuristic terms is located, in domains of similar studies in mathematics, physics and biology;
4. In research terms is located at the interface between the bio-psycho and social domains.

5. It reflects the symbolic determinants, at the word-sentence levels (semantics), but it is relatively independent from it.

6.2.2 Rhythms for entrainment

There is a powerful amount of literature on social rhythms and the way they are capable to synchronize social groups. There are studies on commuting and their impact on the productivity of nations and biorhythms of individuals. There are important historical studies on how the choice of the calendar and holydays can shape the identity of religions or nations (see the different choices of Sunday, Saturday and Friday for the three monotheistic religions). At the same time national revolutions have established new social timings defining new calendars: the French and the Russian revolution are the most recent examples (Zerubavel, 1981).

Social timing is a powerful tool for social entrainment. In the same way, to induce synchronization and change, you can use timing as shamans well knew using tribal rhythms and dances (Freeman, in press). Psychotherapies have been trying to do the same, and there is an interesting example on how Freud tried to find the right timing technique to induce entrainment.

Fachinelli (2001) examines the relation between analytical practice and time. He refers to three types of activity found in Freud's work.

- First, his early 'nearly feudal' treatment of hysterics when its duration was contained within the limited period of the illness time, and he went to cure at home.
- Then an 'almost utopian' practice based on a 'friendly relationship', when his conversation with Gustav Mahler during a walk they had therapeutic effects.
- The third 'industrial' practice, drawn from Freud's papers on technique, prescribes the use of a "Tayloristic" Grand Clock to regulate the analytical relationship using the time organization of factories.

While the analytical work thus became chronometrical and fixed, by acquiring organization; the scientific apparatus seemed to necessitate a progressively longer period of time, because its aim became gradually less distinct, indefinite, and even endless. Time, limited and defeated in the present of the session, seems to shift to an unreal victory in the future. Such a monotonous and institutional relationship in analysis ends up existing for itself and by its own rules, without concern for its own end.

Freud's Grand Clock in fact recalls, on the one hand, the Galilean-Newtonian conception of homogeneous time, which can be divided into equal segments. On the other hand, it has a precise economic significance of a proto industrial Tayloristic organization of factories. Its severity tends towards the total elimination of empty, gratuitous moments, which become rejected fragments, and it goes considerably beyond the individual artifice of a "music or language teacher", as he defined his activity. It has greater rigor and inflexibility and in fact has become widespread and accepted in many different "cures of the soul".

It recalls another time: the time of mechanical clocks which in the late 18th century marked the advent of the industrial machine.

The machines were different, but they all worked according to an exactly divided time; *they constituted a replica of the great universal clock of classical mechanics*. These machines, by means of high synchronization, eliminated the waste of time and sharply increased productivity. The resulting regular and constant work *thus abolished the "characteristic discontinuity" of previous work schemes*. The acceptance of this setting by Freud's followers was unanimous. But, it was in contradiction with the basic linguistic rule of every treatment that in English has had a mistake in translation from the German, revealing a shift in the paradigm.

6.2.3 Einfall

It begins formally at that moment when the analyst invites his traveling companion to "*communicate everything which comes to mind*". The Freudian formula is: "*alles mitzuteilen was [ihm] einfällt*". "That which comes to mind" is in German *was einfällt:* literally, that which falls or descends into the mind. This event is an *Einfall*, a sudden, unexpected incidence or occurrence: a word dating back to the mystics of the late Middle Ages. Every single *Einfall* is connected to successive ones by means of threads (*Fäden*) or lines (*Linien*) or concatenations (*Verkettungen*).

Freud used to complain that his therapeutic outcome were better when he was younger, and his approach was not so rigorous, than after he developed his Grand Clock. This is well documented in a book of *memoirs* from his patients (Albano, 1987), where we find the extraordinary variety of his *real* approaches, and we discover that he was less Freudian we would imagine.

Mostly we find that he was tuned in a different way to the different states of each analysand, and used many different settings and timings. This meant that he was focused on the nonlinearities. *Linear timing means endless periodicities or fixed states: Nash homeostatic equilibriums instead of Self Organized Criticality.*

Not to mention that Post-industrial and Post-modern societies based on the Information Technology rely on social organizations of time very different from the old Grand Clock of factory based economies.

6.3 Play

The tractability of *social play as an evolved phenotype* also makes it attractive to those interested in learning more about nonhuman and human cognition. Understanding activities such as play may also be important for developing new research dealing with comparative approaches to cognition, including building androids.

How important it is to negotiate play and to agree that play is the name of the game? *When animals play, behavior patterns for the most part are "borrowed" from other contexts, and individuals need to be able to tell one another that they do not want to eat, fight with, or mate with the other individual(s), but rather, they want to play with them.*

In most species in which play has been observed, specific actions have evolved that are used to initiate ("I want to play with you") or to maintain ("I still want to play with you regardless of what I just did to you or regardless of what I am going to do to you") play. These actions seem to function in negotiations between participants, the result of which is that they foster an agreement to engage in play rather than to partake in aggression or predation, for example. There is no solid evidence that animals invite others to play and then exploit them (Bekoff, 1998). Furthermore, self-handicapping and role reversals have also been observed, in which, for example, dominant individuals allow themselves to be dominated only in the context of play.

"Play is all motor activity performed afterbirth that appears (our emphasis) to be *purposeless*, in which motor patterns from other contexts may often be used in modified forms and altered temporal sequencing. If the activity is directed toward another living being is called social play."

Note that this is a positive definition, in that play is defined in terms of what it is. For years, it was trendy to refer to social play as being not aggression, not predation, or not reproduction, but these sorts of definitions proved to be of little or no value to those who were interested in learning more about play. The definition centers on the structure of play sequences (what animals do when they play) and not on possible functions of play.

The participants need to cooperate and to negotiate agreements so that they can cooperate in ongoing social play. And, after they agree to play there seem to be social regulations and social obligations to maintain the play mood. Furthermore, there are data that suggest that animals often seek permission to play with another animal, by performing behavior patterns that indicate that play is desired and not another activity.

For example, in the self-handicapping (e.g., inhibited biting or clawing) and role-reversing (being chased rather than chasing) in which dominant animals engage to get more subordinate individuals to play, the dominant individuals are seeking permission to engage in play with animals who might otherwise avoid interacting with them. The end result is that the two individuals agree to cooperate, with self-handicapping and role-reversing being a form of negotiating this agreement. These data are important to consider in discussions of the evolution of cognition for they tell us that certain cognitive abilities must be present even before an individual has the linguistic sophistication of a human.

Because play is composed of actions that are also used in other contexts, an individual needs to be able to communicate to potential play partners that he is not trying to dominate them, eat them, or mate with them. Rather, he is trying to play with them. Behavioral observations of many animals who engage in social play suggest that they desire to do so and believe that their thoughts of the future - how the individuals to whom their intentions are directed would be likely to behave - would be realized if they clearly communicated their desires to play using signals that in some cases seem to have evolved specifically to communicate *play intention.*

While other animals might also cooperate, including those for whom we are hard-pressed to grant the presence of sophisticated cognitive abilities (e.g., cleaning fish), the flexibility of play and the ease with which even two strange animals are able to play after only brief introductions to one another, lead to the conclusion that interactions between players are not merely the result of (somewhat hard-wired and inflexible) coevolved systems. They are the result of ongoing negotiations that require careful assessments throughout an encounter. It seems to be the flexibility of play that is important in allowing animals to negotiate the desired outcome.

How are agreements reached in the context of social play? How is turn-taking by individuals accomplished? In most species in which play has been described, play-soliciting signals appear to foster some sort of cooperation between players so that each responds to the other in a way consistent with play and different from the responses the same actions would elicit in other contexts (Bekoff, 1998); play-soliciting signals provide aid to the interpretation of other signals by the receiver.

Do available data suggest that some animals who engage in social play have a theory of mind: can they and do they make attributions of mental state to others?

Cetaceans and man play synchronized underwater (Colbert, 2002)

6.3.1 Play is SOC

Winnicott (1971; 1971b) proposes that cure happens where there is a meeting between two areas of play, and two persons willing to play together. When play is not possible the therapist's task is to bring the other in a state in which being able to play. It is a state of fusion in which the boundaries between subjective and objective experience can disappear. Psychotherapies are specialized environments to allow people to play. It is important that aggression and sex are outside the door.

Play is probably the activity that in mammals is more similar to the *edge of chaos* fitness landscape: the Self-Organized Criticality. In play you can find a sequence of intentions and cognitions that has the highest number of degrees of freedom, and maximal amount of information. Creativity can be explored and you can meet the Adjacent Possible.

In each session is important to reach the vicinity of play, the letting go and relaxation of a benevolent regression. Long regressions are usually exported out of the protective environment and give the partners the risk of disruptions. So, it is important that the therapist ensures the reestablishment of a closure before the beginning of the play session. To keep alive the therapeutic play the survival of both partners and the environment has to be ensured.

Most of the therapeutic work is to make the other to become able to play, and the task of the therapy becomes the therapy in itself. The main sensor of the real evolution can be the subjective capacity of the therapist to enjoy the meetings and plays. Each encounter and the following separation become sources of any possible threat or joy (Searles, 1979).

6.4 Changing minds

The problem is the phase transition from a series of discrete ideas to a stream of consciousness. Linguistic changes can be brought on by internal dynamics, but also by extra-linguistic events. These may be natural events, such as floods, droughts or volcanic eruptions, or socio-political events, such as the emergence of an aggressive political or religious group, or some striking technical innovation, or simply an entry into new and pristine territory (Hagege, 1993; Dixon, 1998). This relativistic notation is directly related to the idea that living systems are open. We know how in biology change can be related to interaction and symbiosis.

We can see magnified, at the macro level, some effects of inter-subjective couplings. Linguistic change during synchronization is usually evident at the semantic level in phenomena called *pidginization* and *creolization*.

A pidgin refers to a *contact language* used by two groups, one being socially dominant but neither speaking it as a first language. There is a social or cultural distance marked by the necessity to communicate. *Pidgin* language (origin in Engl. word `business'?) is *nobody's native language*; may arise when two speakers of different languages with no common language try to have a makeshift conversation. *Lexicon usually comes from one language, structure often from the other.* But the pidgin is the result of the creation of a new (relatively simple) linguistic system. Because of colonialism, slavery etc. the prestige of Pidgin languages is very low. *Many pidgins are 'contact vernaculars', may only exist for one speech event.*

Creole (orig. person of European descent born and raised in a tropical colony) is a language that was originally a pidgin but has become nativized, i.e. a community of speakers claims it as their first language. This word next has been used to designate the language(s) of people of Caribbean and African descent in colonial and ex-colonial countries.

Pidgins rely heavily on the relationship between form and meaning, where one form corresponds to only one meaning. Other characteristics of pidgins include:

It has also been shown that the speech of Language2 learners is often reduced in similar ways, for example by leaving out grammatical morphemes and simplifying negation. An attempt has been made in recent years to link the similarities between the reduced language of L2 learners and that of other simplified forms, including creoles and pidgins. Pidgins evolve much like inter-languages, changing from one point in time to the next, so the development of an L2 learner's inter-language may be governed by many of the same rules as a pidgin language system. Some process of linguistic pidginization should be ubiquitous in the entrainment between linguistic subjects. This has been called "convergence through reduction to minimal linguistic universals" (Hjelmslev, 1970 & 1971). This means that initial transitions to cognitive synchronization may transit in states of simplified dynamics (creodes).

Pidginization and creolization are the linguistic equivalents of coupling and synchronization at the macro level of social interactions between different systems.

6.5 Origins of sex

If at the semantic level we find the cultural processes of pidgin and creole languages, at the informational level the analog of the examined phenomena is found in the origins of sex:

"...in our sexual activity we are not only thermodynamically but informationally open...in order to continue evolutionarily we open ourselves to fresh genes." (Margulis, 1986)

Bacteria living inside each other shared each other's food, metabolites and eventually genes. Meiotic sex in the reproduction of animals and plants is just a part of sex. Sex is a genetic mixing process that is not reproduction as we see in mammals.

Sex is the formation of a genetically new individual that contains genes from more than a single source, and it is always happening in synchronization and information exchange. Reproduction is an increase of number of individuals, and doesn't necessarily imply sex.

In the Acrasia cellular slime molds hundreds of individual cells independently living, come together under the influence of a chemical attractant. They form larger creatures, slugs, but when they separate again there is no increase of the cell number, but there has been sex (information exchange).

In this sense, we could say that everyday life conversations, meetings etc. are sexual activities, because always include information exchanges. They are not necessarily reproductive activities even if they allow courtship and mating. Although additional living beings are often reproduced by a contribution of genes from more than a single parent, sex in most organisms is still divorced from growth and reproduction that are done by nonsexual means.

"The private activities of early cells are involved even today in courtship among human beings" (Margulis): mammalian sex is a late and special variation. Human sexual pleasure has more to do with reproduction than with sex. The pursuit of love and sexual pleasure is the mammalian prelude to reproduction. Nucleated (eukaryotic) cells began as bacterial (prokaryotic) symbioses and are legacies of coevolved microbial couplings. Individuals are aggregate of composite structures.

6.6 Windows of time

St. Augustine's in the Confessions (specifically book II) writes on the paradoxes of "a past-containing nowness" (*distentio animi*). More recently, the pragmatic psychology of William James as found in the famous Principles of Psychology (1920), proposes the definition of *specious present.*

Time in experience is quite different from time as measured by a clock. Time in experience presents itself not only as linear but also as having a *complex texture* (evidence that we are not dealing with a "knife-edge" present) a texture that dominates our existence to an important degree (Varela, in Petitot, 1999).

We could define three levels of temporality:

1. A first level proper to temporal objects and events in the world. This level is close to the ordinary notions of temporality in human experience, which grounds the notion of temporality currently used in physics and computation.

2. A second level, which is quickly derived from the first by reduction, the level of acts of consciousness that constitute object-events. This is the "immanent" level, the level based on the "internal time" of acts of consciousness. Their structure forms the main body of the phenomenological analysis in Husserl's Lectures.

3. Finally (and this is the most subtle level of analysis), these first two levels are constituted from another level where no internal-external distinction is possible, and which Husserl calls the "absolute time constituting flow of consciousness".

As phenomenological research itself has repeatedly emphasized, *perception is based in the active interdependence of sensation and movement.* Several traditions in cognitive research have, as well and in their own way, identified the link between perception and action as key. *It is this active side of perception that gives temporality its roots in living itself.* Within this general framework, we will concentrate more precisely on the structural basis and consequences of this *sensory-motor integration* for our understanding of temporality.

This overall approach to cognition is based on *situated embodied agents*. Varela (1991; 2001) has introduced the name *enactive* to designate this approach more precisely. It comprises two complementary aspects: (I) the ongoing coupling of the cognitive agent, a permanent coping that is fundamentally mediated by sensory-motor activities;

(2) the autonomous activities of the agent whose identity is based on emerging, endogenous configurations (or self-organizing patterns) of neuronal activity.

Enaction implies that sensory-motor coupling modulates, but does not determine, an ongoing endogenous activity that it configures into meaningful world items in an unceasing flow.

From an enactive viewpoint, any mental act is characterized by the concurrent participation of several functionally distinct and topographically distributed regions of the brain and their sensory-motor embodiment. From the point of view of the neuroscientist, it is the complex task of relating and integrating these different components that is at the root of temporality.

These various components require *a frame or window of simultaneity that corresponds to the duration of lived present*. There are three scales of duration to understand the temporal horizon just introduced:

- basic or elementary events (the "1/10" scale);
- relaxation time for large-scale integration (the "1" scale);
- descriptive-narrative assessments (the "10" scale).

The first level is already evident in the so-called *fusion interval* of various sensory systems: the minimum distance needed for two stimuli to be perceived as non-simultaneous, a threshold that varies with each sensory modality. These thresholds can be grounded in the intrinsic cellular rhythms of neuronal discharges, and in the temporal summation capacities of synaptic integration. These events fall within a range of 10 msec (e.g., the rhythms of bursting inter-neurons) to 100 msec (e.g., the duration of an EPSP/IPSP sequence in a cortical pyramidal neuron). These values are the basis for the 1/10 scale.

Behaviorally, these elementary events give rise to *micro-cognitive phenomena* variously studied as perceptual moments, central oscillations, iconic memory, excitability cycles, and subjective time quanta. For instance, under minimum stationary conditions, reaction time or oculomotor behavior displays a multimodal distribution with a 30-40 msec distance between peaks; in average daylight, apparent motion (or "psi-phenomenon") requires 100 msec.

This leads naturally to the second scale, that of long-range integration. Component processes already have a short duration, on the order of 30-100 msec; how can such experimental psychological and neurobiological results be understood at the level of a fully constituted, normal cognitive operation? A long-standing tradition in neuroscience looks at the neuronal bases of cognitive acts (perception-action, memory, motivation, and the like) in terms of cell assemblies or, synonymously, neuronal ensembles. A cell assembly (CA) is a distributed subset of neurons with strong reciprocal connections.

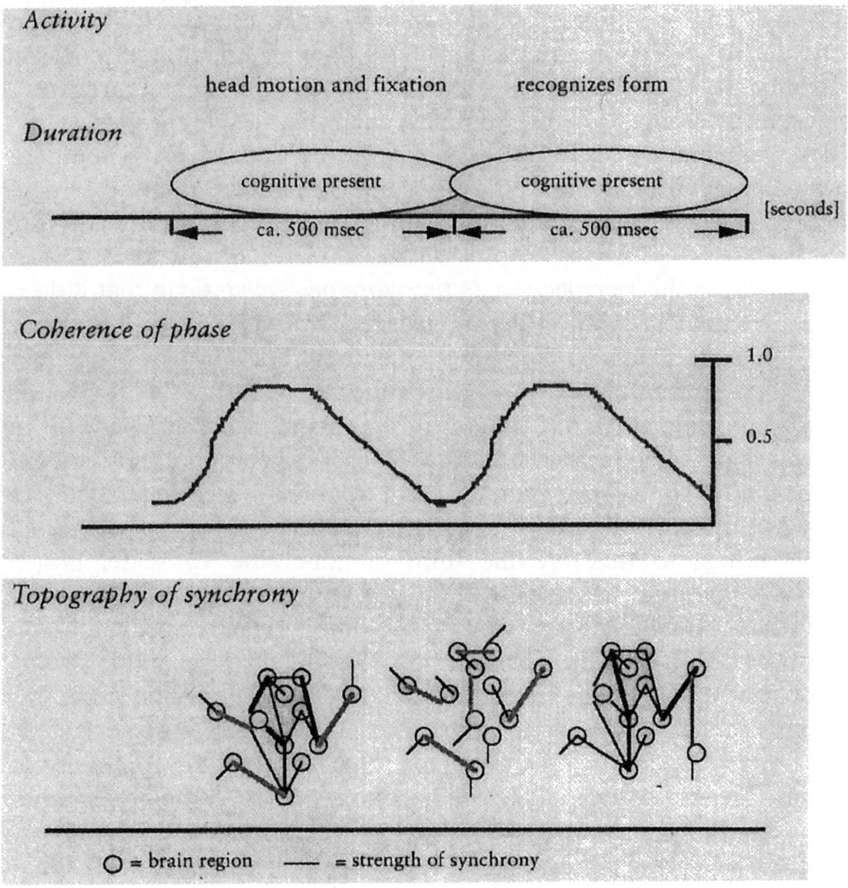

Windows of time (Varela, 1991)

The diagram depicts the three main hypotheses. A cognitive activity (such as head turning) takes place within a relatively incompressible duration, a *cognitive present*. The basis for this emergent behavior is the recruitment of widely distributed neuronal ensembles through increased frequently, coherence in the gamma (30-80 Hz) band. Thus, the corresponding neural correlates of a cognitive act can be depicted as a synchronous neural hypergraph of brain regions undergoing bifurcations of phase transitions from one cognitive present content to another.

Recently this view has been supported by widespread findings of oscillations and synchronies in the gamma range (30-80 Hz) in neuronal groups during perceptual tasks. Thus, we have neuronal-level constitutive events that have a duration on the 1/10 scale, forming aggregates that manifest as incompressible but complete cognitive acts on the 1 scale.

This completion time is dynamically dependent on a number of dispersed assemblies and not on a fixed integration period; in other words it is the basis of the origin of duration without an external or internally ticking clock.

Nowness, in this perspective, *is therefore pre-semantic* in that it does not require a re-memoration in order to emerge. The evidence for this important conclusion comes, again, from many sources. For instance, subjects can estimate durations of up to *2-3 seconds* quite precisely, but their performance decreases considerably for longer times; spontaneous speech in many languages is organized such that utterances last 2-3 seconds; short intentional movements (such as self-initiated arm motions) are embedded within windows of this same duration.

This brings to the fore the third duration, the 10 scale, proper to *descriptive-narrative* assessments. In fact, it is quite evident that these endogenous, dynamic horizons can be, in turn, linked together to form a broader temporal horizon. This temporal scale is inseparable from our descriptive-narrative assessments and linked to our linguistic capacities. It constitutes the “narrative center of gravity” in Dennett's metaphor (Dennett 1991), *the flow of time related to personal identity*. It is the continuity of the self that breaks down under intoxication or in pathologies such as schizophrenia or Korsakoff's syndrome. As Husserl points out, commenting on similar reasoning in Brentano: “We could not speak of a temporal succession of tones if … what is earlier would have vanished without a trace and only what is momentarily sensed would be given to our apprehension” To the appearance of the just-now one correlates two modes of understanding and examination (in other words, valid forms of donation in the phenomenological sense):

(I) remembrance or evocative memory and (2) mental imagery and fantasy.

The *Ur-impression* is the proper mode of the now, or in other words, it is where the new appears; impression intends the new. Briefly: *impression is always presentational, while memory or evocation is re-presentational.*

This behavior embodies the important *role of order parameters in dynamical accounts.*

Order parameters can be described under two main aspects:

I. The current state of the oscillators and their coupling, or initial conditions; and

2. The boundary conditions that shape the action at the global level: the contextual setting of the task performed, and the independent modulations arising from the contextual setting where the action occurs (namely, new stimuli or endogenous changes in motivation).

6.7 Doubling

Around 1911 Husserl introduced the term *double intentionality* (1964) for this articulation, since there is not only retention (of the object-event) but also retention of retention (a reflective awareness of that experience). James came to a similar conclusion when he described his experience of lying in bed, telling himself to get up. He stayed in bed. He later found himself up but could not remember the exact instant of transition. Libet (1993; 1999) experiments on the consciousness lag, though discussed in their methodology, suggest the same solution: there is a doubling between intentionality and consciousness, and it has neurobiological foundations.

Vitiello (2001), starting from the "quantum brain" hypothesis (whose problems we won't discuss here) suggests: "In the dissipative quantum model of brain memory recording is modelled as coherent condensation of certain quanta in the brain ground state. The formation of finite size correlated domains allows the organization of stored information into hierarchical structures according to the different life-times of memories and the size of the corresponding domains. The openness of the brain to the external world (dissipation) *implies the doubling of the brain system degrees of freedom*. The system obtained by doubling, the *Double, plays the role of the bath or environment in which the brain is permanently embedded. It is suggested that conscious as well as unconscious activity may find its root in the permanent dialogue of the brain with its Double."* (Italics are mine).

Jacques Lacan (1977) schematized this dialectics in the *mirror phase* of infant development: the birth of the subject's identity in front of the mirror is also the moment in which the *Ur-spaltung* between the subject of intentionality, the Es, and the subject of consciousness, the Ego, crystallizes. This is the "*I scheme*" that we obtained from a deep modification of the Lacan's so-called L scheme:

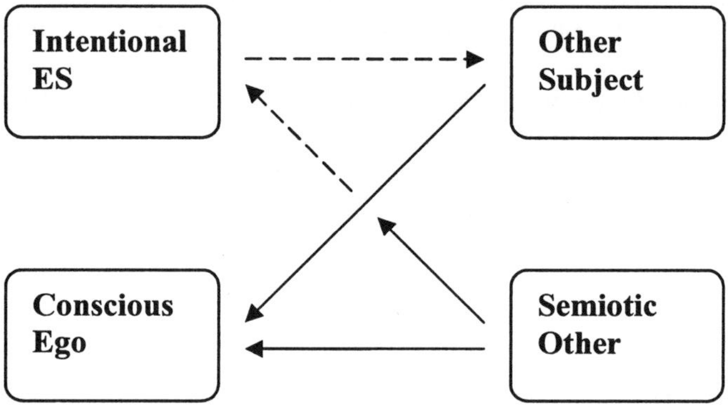

This scheme of the *dynamics in subjective experience* offers the crossing of the two main vectors: the ***vector of the imaginary***, between OS and the Ego; and the ***vector of the unconscious*** experiences, between the (Semiotic) Other and the Es.

6.7.1 Neurodynamics of time

The neurodynamics of time we have been pursuing is essentially based on nonlinear coupled oscillators. As we saw, this class of dynamical systems finds its wealth of behavior in the fact that *constitutional instabilities are the norm* and not a nuisance to be avoided (Varela, in Petitot, 1999). The case of multi-stability makes this quite evident experientially: the percepts flip from one to another (depending on order parameters) by the very nature of the geometry of the phase space and the trajectories.

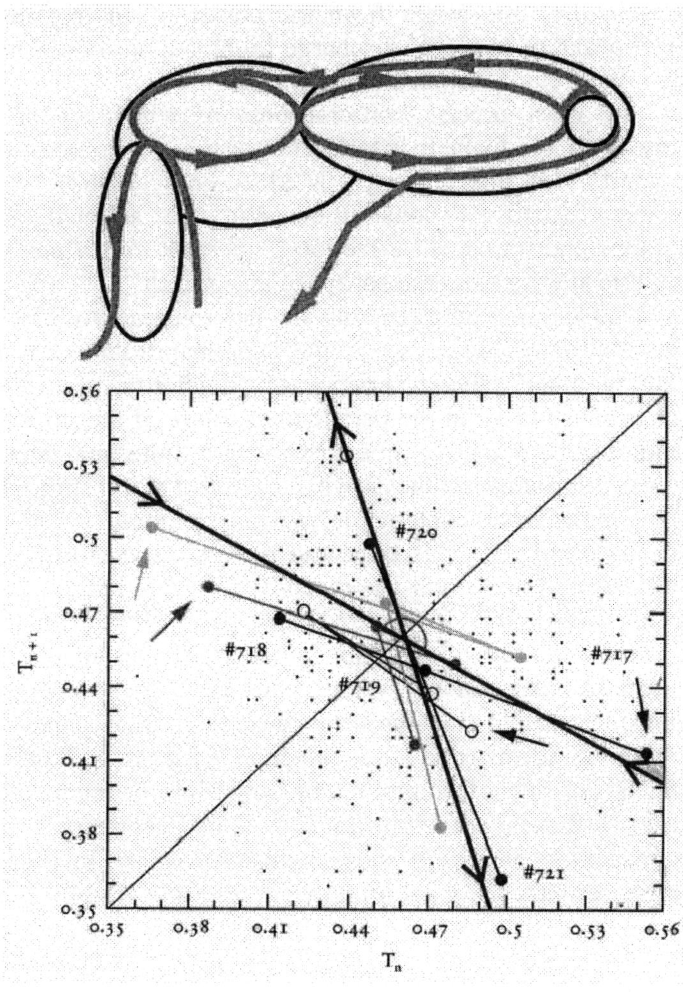

Multiple instabilities (Varela, in Petitot, 1999)

Top, a highly schematic diagram of the current view that a complex, chaotic dynamic should be regarded as having a geometry in phase space where multiple instabilities are found locally (gray lines). A system's trajectory shifts (in black lines) constantly from one local instability to another, in an unceasing flow, under the modulation of boundary conditions and initial conditions. Bottom, experimental evidence for local saddle instabilities in the time series from the cortex of an implanted epileptic patient. The peaks of the local discharge in the temporo-occipital cortex are followed in their return map or Poincaré section.

"Even the most precise consciousness of which we are capable is affected by itself or given to itself. The very word consciousness has no meaning apart from this duality," (Merleau-Ponty, 1962). Alterity is the primary clue for time's constitution. We are affected not only by representations and immanent affection ("*affection de soi par soi*"), but by the inseparable alterity from the sphere of an *ego-self.*

The very distinction between auto and hetero ceases to be relevant, since in all cases it all is brought down to the same manifestation: it is a question of "something other", the experience of an alterity, a difference in the identity of is constitutive in the paradoxical nature of the shifter called Ego. Subject and object at the same time, or with an infinitesimal delay.

We seek a non-dual synthesis whereby *affect* is constitutive of the self and at the same time contains a radical openness or unexpectedness concerning its occurring. This bootstrap principle seems to be present in a variety of natural systems and has recently been referred to as *operating at the edge of chaos*, or *self-organized criticality*. For example, this idea provides a renewed view of evolution, since it provides an answer to the old nature (genetic expression) versus nurture (environmental coupling conditions) dilemma. In this synthetic view (Kauffman, 1993) the relation between natural forms and the selection process in their ecological embedding is not one of contradiction but precisely mutual imbrications. This built-in shiftiness, enfolding trajectories and geometry, gives a natural system a possibility of always staying close to regions in phase space that have multiple resources (e.g., at least two in bi-stable visual perception).

6.7.2 Intentional couplings

All that brains can know is the hypotheses they construct and the results of testing them by acting into the environment, and *learning by assimilation from the sensory consequences of their actions*. The process is described as *intentionality*. It works through the action-perception-assimilation cycle. The cost of this solution to the problem of infinite complexity by hypothesis testing is the progressive isolation of individuals, as they accumulate their unique experiences through which their personalities form. Socialization and the acquisition of shared knowledge requires the emergence of new personality structure by self-organization through *chaotic dissolution* of existing structure, as prelude to creation of new traits, habits and values. *Dissolution work is a crisis situation by regression to earlier stages of development, from which a fresh start can be made.*

Socialization requires neuro-chemical mechanisms of affiliation and bonding that evolved through the requirements of parental care of altricial offspring in mammalian reproduction. These mechanisms are invoked by means of behavioral techniques from cultural evolution. The dynamics, neural mechanisms, behavioral signs, methods of induction, and therapeutic utility of dissolution should be known by therapists. Lack of recognition and understanding may cause failure to use brief windows of opportunity to instill long-term relief of psychic pain by restructuring intentionality in distressed patients.

Freud (1985) in his attempt to lay a scientific foundation for psychoanalysis, later abandoned as "premature", labeled nerve energy as "Q" and wrote:

"This line of approach is derived directly from pathological clinical observations, especially those concerned with excessively intense ideas. These occur in hysteria and obsessional neurosis, where, as we shall see, the quantitative characteristic emerges more plainly than in the normal. ... What I have in mind is the principle of neuronal inertia, which asserts that neurons tend to divest themselves of quantity (Q). ... We arrive at the idea of a cathected neuron (N) filled with a certain quantity [of energy]. ... The principle of inertia finds expression in the hypothesis of a current, passing from the cell-processes or dendrites to the axon. ... The secondary function [memory] is made possible by supposing that there are resistances which oppose discharge ... in the contacts [between the neurons] which thus function as barriers. The hypothesis of 'contact-barriers' is fruitful in many directions."

Indeed that was an understatement, in that the greater part of biological psychiatry is now based on synaptic transmitter chemistry.

6.7.3 From energy to information

In the 20th century the core concept of nerve energy was replaced by 'information', which was carried by action potentials and processed in networks of neurons in accordance with the parent science of information theory, *'embodied cognition'*, in which the body serves as the principal tool of the brain for cognitive development and the acquisition of knowledge, and as *'situated cognition'*, in which the structure of the environment determines, and is determined by, the actions of individuals seeking knowledge. All knowledge gained by individuals is earned by their actions into the world.

Assimilation is not adaptation by passive information processing, nor is it an accumulation of representations by resonances. It is the shaping of the self to bring it into optimal interaction with desired aspects of the world.

The goal of an action is a state of competence that Maurice Merleau-Ponty (1945) called *maximum grip*. Assimilation is the beginning for all knowledge. Thus the manner of acquisition of knowledge is by thrusting the body into the world, from which our word 'intention' has come from the Latin *intendere* = 'stretching forth'. The thrust initiates the action-perception cycle, which is followed by the changes through which the self learns about the world by assimilation (from the Latin *adaequatio* = adequacy) of the self to the world. There is no transfer of information across the senses into the brain, but instead the creation of information within the brain under the existing constraints of the brain and body.

In this respect cognition is related to *digestion*, which protects the integrity of the immunological self by breaking all forms of food stuffs into elementary ions and molecules that are absorbed and built into complex macromolecules, each now bearing the immunological signature of the individual self. Similarly experience are digested, events and objects in the world are broken into sheets of action potentials like pinpoints of light, the *raw sense data* of analytic philosophers and the phantasms of Thomists, and new forms emerge through constructions by the chaotic dynamics in sensory cortices. The explanation for this manner of function of both the neural and the digestive systems is essentially the same: the world is infinitely complex and the self can only know and incorporate what the brain makes within itself. This is why neurobiologists using passive neural networks cannot solve the figure-ground problem, why linguists cannot do machine translation, why philosophers cannot solve the symbol grounding problem, why cognitive scientists cannot surmount the limitations of expert systems, and why engineers cannot yet build autonomous robots capable of operating in unstructured environments.

6.8 Presentation *vs.* representation

Every sensory cortex maintains a landscape of chaotic attractors that correspond to the perceptual categories that the subject is capable of discriminating (Ohl, Scheich and Freeman 2001). When the subject attends to an expected event, the landscape is created, and a known stimulus gives access to the proper basin. After the dynamics converges to the selected attractor, the specific details of the stimulus are removed. The raw sense data by which category selection is performed are discarded after they are no longer useful. These properties of chaos are most in evidence when a hypothesis is disproved by testing, because the resulting stimulus is novel. Upon this failure, the prediction is changed by creative activity in the limbic system and disseminated by corollary discharges, so that a new hypothesis is tested.

The process is repeated by the creation of new attractors by trial and error, until a reward is forthcoming and a hypothesis is proved. The changes generated by chaotic dynamics are in the form of the modification of numerous synapses that interconnect the neurons in many parts of the brain. When the process finally succeeds, the changes bring about assimilation not by the incorporation of the forms or the information offered by the world but, instead, by a creative reshaping of the brain as well as the body that facilitates continuing interactions of the self with its world, in so far as that world is accessible to the brain and body. Learning to dance, to play the violin, or to play tennis *requires changes in the body as extensively as in the synapses of the brain*. In this view consciousness comes only after an action has been initiated, not before, so that the action is perceived by the actor as a cause, and the sensory consequences as effects. This self-awareness of the intent-action-perception-assimilation cycle is the basis for the concept of causation and for the critical importance of the time lag between effect and its necessarily prior cause.

Studies by Piaget (1980) have shown that this association of cause and effect is laid down in the somato-motor phase of development, when an infant is learning to control its body. How and why the experience of awareness comes soon after in the way that happens are matters for speculation, but its phenomenology gives clear indication of both the powers and the limitations of this remarkable process. Its temporal range is enormous, but its momentary content is sparse. Our actions are influenced by our entire life history, but we are conscious of only minute, intermittent fragments in sequential moments of thought.

There is compelling clinical and physiological evidence that the medial temporal lobe is essential for the construction of the life history, which gives the wholeness of intentional structure, the self, personhood, or, more popularly, personality. However, that brain part is not necessary for consciousness, nor is the hippocampus a site of long term memory storage. On the contrary, the dynamic processes that comprise intentionality in action are non-local, and likewise the personality to which it gives rise.

Furthermore, *most of intention is inaccessible to consciousness at any one moment, not so much that access is forbidden, rather that the sheer massiveness of the dynamic state space precludes squeezing the entire past through the eye of a needle. Reification of intentionality as the Unconscious appears to reduce this living body of experience inappropriately to a static warehouse of latent memories with locked doors. New developments in the sciences of nonlinear dynamics, complexity and chaos indicate that this conflict is a pseudo-problem* (Freeman 2001).

6.8.1 Forget to learn

No human action is entirely determined by genes, world or self but instead by an ever-shifting balance of interactions among the three formants. The challenge for neuroscientists is to analyze the brain dynamics by which goals emerge in the form of imagined future states, which lead to neural patterns of selective attention, pre-configuration of expected input, and actions that are intended to bring into conformance the flow of sensory input with the desired state.

How do we come to dream of that which should be but is not, and then act to realize the dream? Assimilation through Hebbian learning is not enough to explain the neurobiological mechanisms by which understanding develops that bonds individuals by the process of socialization. Synaptic learning unchecked leads to isolation, lack of empathy, and inability to act in concert. Examples of extreme failures of socialization may include some autistic children and psychopaths who have astonishing cognitive skills but stunted emotional development. The biological techniques for inducing *dissolution* are well known. Individuals separate themselves or are isolated from their normal social surroundings and support systems. They engage in or are subjected to severe physical exercise as in dancing, sports, and military drills, lack of sleep, and chemical stresses of their brains through purgatives and fasting, and the induction of powerful emotional states of love, hate, fear or anger.

At some threshold the customary structure of the individual begin to crumble, and a collapse may occur that was described by Ivan Pavlov as *transmarginal inhibition*, the stage of physiological arousal beyond which further excitation leads to paradoxical depression. The experience may range from ecstatic visions of angels and blinding illumination through degrees of elation or discomfort to the stark terror of psychic free fall. There is *regression* to successively earlier levels of assimilation as the structure of intentionality dissolves, particularly with resurfacing of old patterns of relations to parental care. There is a loss of normal constraints on behavior, and, in extreme instances, of language, locomotion, posture and even consciousness as the individual collapses.

Recovery from collapse is followed by a state of extreme suggestibility, in which the skills of language and the competencies of daily living are regained, yet new values and habits can be established. This is done in a social setting of succor and loving care by attendants who induce by example and exhortation the cooperative behaviors that lead to shared beliefs and, above all, to blind trust in the new companions and the social organization they embody and provide. A state of malleability emerges in the depth of crisis, in which compassionate companions through loving care can invite cooperative actions. Joint actions support the growth of a new life style based in trust.

This is a *two-way process*, because the caregivers get strong feelings of satisfaction from their supportive actions, and the recipients have strong sensitivity to peer pressures experienced as feelings of need for approval. The process is frequently referred to as being reborn. In the absence of support there is re-establishment of the status quo ante, meaning that the opportunity for change can be lost, attesting to the high degree of dynamic stability that characterizes intentional structures in normal circumstances.

Little is known about the neurophysiology of these personal and social interactions and even less about the neurochemistry of the changes in brains that occur in and are induced by dissolution. The evolutionary antecedents of the techniques for socialization are likely to be found in the processes by which *bonding* occurs in altricial mammals as a necessary basis for reproduction (Freeman 1995).

The most likely candidate for a leading role in dissolution is a chemical neuro-modulator named *oxytocin*. This neuropeptide has been known for many years as the agent in the female body that induces labor in parturition and subsequently lactation in nourishment of the young. More recently oxytocin has been found to be released by the brain into itself during sexual intercourse, particularly during orgasm in both men and women, and to be implicated in pair bonding not only of the parents to the child but also of parents to each other. Sexual orgasm is the basis for pair bonding and cooperation for the nurture of the young. The subjective and objective manifestation of the madness (*folie á deux*) that accompanies the dissolution of intentionality in humans is wonderfully described in "Troilus and Cressida". An excessive concern for computational algorithms currently leads many scientists to neglect these non-computational processes as the bases for pair bonding and group behavior.

The neurochemical actions of oxytocin in the brain are widespread, extremely complex, and difficult to study, so that much remains to be explored, but present knowledge shows that this neuropeptide is capable of *inducing the meltdown of past learning that enables new learning*. A simple example is the release of oxytocin flooding the brain of the multiparous ewe during delivery of her second and later litters, following which the dam refuses to nurse her earlier litters, having expunged the olfactory imprint required for maternal recognition of them as her offspring. This primitive but well documented instance of dissolution serves also to explain its biological utility. Oxytocin is not likely to act alone, rather *in concert with other neuropeptides*, the neuroamines, and an array of amino acids from the brain stem nuclei and periaqueductal gray matter, all known to mediate states of emotion and levels of affect and disposition. However, existing data and theory are alike inadequate to the task of modeling neurochemical systems of this complexity. Relations between oxytocin and nerve growth factor should be investigated, because of the implications of NGF in growth and pruning of neurons related to learning from experience (Cirulli, 2001).

6.9 Regression

Freeman postulates that affiliation is realized through new learning by cooperative behaviors driven by brains that have been prepared by the neurochemical changes precipitating dissolution, a *state transition* that leads to regression and clears the way to formation of new brain circuits. Cooperative action is the bedrock of social bonding for the same reason that brains work by creating and testing hypotheses as their means for information processing. The *social technology of bonding* is well known, having been explored by anthropologists in studies of tribal rites of passage, ordeals, and ceremonies (Verger 1954), invariably accompanied by use of *timing-devices* like music, drumming, dance, and other forms of predictable repetitive actions, and by *marking-devices* such as flags, icons, totems, and, in modern times, corporate logos, military insignias, and the colored armbands of teenage gangs.

Dissolution may be occurring episodically throughout infancy and childhood and perhaps in minimal degree every night during sleep and dreaming. In these situations we pass through transitions and initiations.

We may infer that the desired changes in brain dynamics of learning take place through Hebbian association and anti-Hebbian habituation, extinction, and normalization, which are enacted and supported by neuro-humoral agents, including the biogenic amines and neuro-peptides in collaboration with nerve growth factors that operate by mobilizing sections of the genome.

Faith healers and shamans in technologically less complex social groupings rely on techniques. There is heavy emphasis on the familial and social contexts to which behaviors are to be adapted, and there is strong reliance on altered states of consciousness involving the induction of strong emotional arousal leading to trance, transmarginal inhibition, and hallucinatory experiences. Psychoanalysis and other psychotherapies have *different mixes of many of these components*. The classic approach using free association may be likened to the unsupervised learning used in recurrent neural networks, the blind search for unknown forms using techniques that are described as random walks, Markov processes, and genetic algorithms. These are the techniques used by bacteria and genes that have no brains.

The concept of transference clearly relates to the dynamics of assimilation, whether to objects, events, situations or other persons, but particularly between patient and therapist. Neuro-dynamics may illuminate the process by enabling the distinction between structural advancement by conventional, constructive learning in Hebbian association, habituation and normalization, versus advancement through discontinuous and illogical modification by chaotic dissolution and rebuilding in new directions.

The distinction is more than a matter of degree of change. At one pole is the cool detachment of behavioral therapy, at the other pole is fierce emotional involvement in abreactive techniques. Different techniques can be effective in different situation, but the general methodology and the theoretical background are ready to be unified. Neurodynamics helps to show that assimilation and dissolution may be active at both extremes and in varying degrees between the poles, while emphasizing those two very different kinds of chemistry, neural processing, and therapeutic methodologies are at work, whether simultaneously or sequentially.

Every patient comes to a therapist after making an intentional decision, which carries an expectation and hope of relief to be found through guidance and benefaction. Likewise the therapist accepts a new patient with the hope of a perfect match and a promise of new explorations on complex and new landscapes.

6.10 Progression

May a last question disturb the small semiotic universe of this book? This last question is: "changing minds is a matter of control, synchronization or co-evolution"? Did we change mind during the travel?

Ovidius, the ancient Roman poet, was asking himself in his *Ars Amandi* about any possible technique to manage love: "..*quas possit Amor remanere per artes…tam vasto pervagus orbe puer….*" ("…how could we manage Love by arts…this child in this immense world…")? He was obviously skeptic about any possible result in our efforts in taming love and all other passions.

Von Neumann, instead, was really confident in the positive resources of technological progress: "All stable processes, we shall predict. All unstable processes, we shall control." (John von Neumann, interview, circa 1950). But, as we know, the total control of highly complex systems can be a dangerous illusion.

So, we will continue to explore the Garden of Forking Paths designed by Borges (1964) in one of his seminal writings: "Time forks perpetually towards innumerable futures. I leave to the various futures (not to all) my garden of forking paths."

In this *promenade* we realistically rely on the wisdom of the Zen monk who wrote this *koan* on co-evolution:

"…trees create the bodily form of the wind…"

Bibliography

Abarbanel HD, Carroll TA, Pecora LM, Sidorowich JJ, and Tsimring LS (1994) Predicting physical variables in time-delay embedding. *Physical Review E.* 49 (3):1840-1853.

Abraham FD, Abraham R, and Shaw CD (1990) *A visual introduction to dynamical systems theory for psychology*, Santa Cruz, Cal: Aerial Press.

Ainsworth M, Blehar M, Waters E & Wall S (1978) *Patterns of Attachment.* Hillsdale NJ: Erlbaum.

Albano L (ed.) (1987) *Il divano di Freud.* Parma: Pratiche Editrice.

Aquinas Thomas (1962) *Philosophical texts*. London, New York: Oxford University Press.

Aristotle (1994) *De anima* (Critical ed.) Leiden: E.J. Brill.

Aristotle & McKeon RP (1941) *The basic works of Aristotle.* New York: Random House.

Augustine (2001) *The confessions*, New York: A.A. Knopf.

Baas NA (19949 Emergence, hierarchies, and hyperstructures. CG Langton, ed. in: *Artificial Life III, Santa Fe Studies in the Sciences of Complexity, Proc. Volume XVII.* Redwood City, Cal: Addison-Wesley.

Babloyantz A and Destexhe A (1986) Low-dimensional chaos in an instance of epilepsy. *Proc. Natl. Acad. Sci. U.S.A* 83 (10): 3513-3517.

Bak P (1996) *How nature works: the science of self-organized criticality*. New York, NY, USA: Copernicus.

Bartholomew K & Horowitz LM (1991) “Attachment styles among young adults: a test of a four-category model,” *Journal of Personality and Social Psychology* 61, 226-244.

Bates E (1976) *Language and context, the acquisition of pragmatics*. New York: Academic Press.

Bateson G (1979). *Mind and nature, a necessary unity*. (1st ed ed.) New York: Dutton.

Bekoff M and Byers JA (1998). *Animal play: evolutionary, comparative, and ecological perspectives*, Cambridge, U.K:Cambridge University Press,.

Berman A, Neumann M & Stern R (1989) *Non-negative Matrices in Dynamic Systems* (Wiley, New York).

Bion WR & Bion F (1992). *Cogitations*. London: Karnac Books.

Borges JL (1964) *Labyrinths. Selected stories & other writings*, New York: New Directions Pub. Corp, 1964.

Bowlby J (1969) *Attachment and Loss: Attachment.* New York: Basic Books, Vol. 1.

Bowlby J (1973) *Attachment and Loss: Separation.* New York: Basic Books,Vol. 2.

Bowlby J (1980) *Attachment and Loss: Loss, Sadness and Depression.* New York: Basic Books, Vol. 3.

Bowlby J (1978) Attachment theory and its therapeutic implications. *Adolesc.Psychiatry, 6,* 5-33.

Bowlby J (1982) Attachment and loss: retrospect and prospect. *Am. J. Orthopsychiatry, 52,* 664-678.

Breuer J & Freud S (1895) *Studien uber Hysterie.* Leipzig: F. Deuticke.

Carroll TL and Pecora LM (1993) Using chaos to keep period-multiplied systems in phase. *Physical Review E.* 48 (4):2426-2436.

Champneys AR & Kuznetsov YA (1994) "Numerical detection and continuation of codimension-two homoclinic bifurcations," *International Journal of Bifurcation and Chaos* **4,** 7 85-822.

Churchland PM & netLibrary, I. (1995). *The engine of reason, the seat of the soul a philosophical journey into the brain.* Cambridge, Mass: MIT Press.

Cirulli F (2001) The role of environmental factors on brain development and nerve growth factor expression. *Physiology & Behavior* 73,.

Cornelissen G (1994) *Introduction to chronobiology*, Minneapolis, Mn: Medtronic.

Cornford F (1957) *From religion to philosophy, a study in the origins of western speculation.* New York: Harper.

Collins NL & Read SJ (1990) "Adult attachment, working models and relationship quality in dating couples," *Journal of Personality and Social Psychology* 58, 644-663.

Darwin C (1977) *The collected papers of Charles Darwin.* Chicago: University of Chicago Press.

Dawkins R (1976) *The selfish gene*, New York, Oxford Univ. Press.

Dennett DC (1978). *Brainstorms. Philosophical essays on mind and psychology.* (1st ed ed.) Montgomery, Vt : Bradford Books.

Dennett DC (1991). *Consciousness explained*, Boston:Little, Brown and Co,.

Descartes R (1998) *Meditations and other metaphysical writings.* London: Penguin.

Diels H & Freeman K (1948) *Ancilla to the pre-Socratic philosophers.* Cambridge, Mass: Harvard University Press.

Dixon RMW (1997) *The rise and fall of languages*, Cambridge UK: Cambridge Univ. Press.

Dubé JL and Desprès P (2000) The Control of Dynamical Systems - Recovering Order from Chaos. In: *The Physics of Electronic and Atomic Collisions,* edited by Y Itikawa, Woodbury,N.Y.:AIP.

Eco U (1976) *A theory of semiotics.* Bloomington: Indiana University Press.

Eliade M (1968) *Le Chamanisme et les techniques archaïques de l'extase, 2e édition revue et augmentée.* Paris: Payot.

Ellenberger HF (1970) *The discovery of the unconscious, the history and evolution of dynamic psychiatry.* New York: Basic Books.

Ellenberger HF (1993) *Beyond the unconscious, essays of Henri F. Ellenberger in the history of psychiatry*. Princeton, N.J: Princeton University Press.

Emde RN, Polak PR, and Spitz RA (1965) Anaclitic depression in an infant raised in an institution. *J. Am. Acad. Child Psychiatry* 4 (4):545-553.

Engel GL & Schmale AH (1972). Conservation-withdrawal: a primary regulatory process for organismic homeostasis. *Ciba Found.Symp., 8,* 57-75.

Feeney JA & Noller P (1990) "Attachment style as a predictor of adultic romantic relationships," *Journal of Personality and Social Psychology* 58, 28 1-291.

Feingold A (1990) "Gender differences in effects of physical attractiveness on romantic attraction: a comparison across five research paradigms," *Journal of Personality and Social Psychology* 59, 981-993.

Fodor JA (1975) *The language of thought*. Cambridge, Mass: Harvard University Press.

Fonagy P (1999) The Process of Change and the Change of Processes: What Can Change in a 'Good' Analysis'. Keynote Address to the Spring Meeting of Division 39 of the American Psychological Association, New York 16th April 1999, in *APA Proceedings*.

Foucault M (1988) *Technologies of the Self.* Amherst, USA, Massachussets University Press

Freeman WJ (1981) A physiological hypothesis of perception. *Perspect.Biol.Med., 24,* 561-592.

Freeman WJ (1986). Petit mal seizure spikes in olfactory bulb and cortex caused by runaway inhibition after exhaustion of excitation. *Brain Res., 396,* 259-284.

Freeman WJ (1995). *Societies of brains, a study in the neuroscience of love and hate*, Hillsdale, NJ:L. Erlbaum Associates.

Freeman WJ (1998). The neurobiology of multimodal sensory integration. *Integr. Physiol. Behav. Sci., 33,* 124-129.

Freeman WJ (1999 and 2001). *How brains make up their minds*. London: Weidenfeld & Nicolson.

Freeman WJ (2000) Mesoscopic neurodynamics: from neuron to brain. *J.Physiol Paris* 94 (5-6):303-322,.

Freeman WJ and Rogers LJ (2002). Fine temporal resolution of analytic phase reveals episodic synchronization by state transitions in gamma EEGs. *J.Neurophysiol.* 87 (2):937-945.

Freud S, Masson JM, and Fliess W (1985) *The complete letters of Sigmund Freud to Wilhelm Fliess, 1887-1904*, Cambridge, Mass: Belknap Press of Harvard University Press.

Freud S (1924) *Collected papers*, New York: The International Psycho-analytical Press.

Fuller TL & Fincham FD (1995) "Attachment style in married couples: relation to current marital functioning, stability over time, and method of assessment," *Personal Relationships* 2, 17-34.

Gabora L (1998) Autocatalytic Closure in a Cognitive System. *Psycoloquy* 9(67).

Griffin DW & Bartholomew K (1994) "Models of the self and other. Fundamental dimensions underlying measures of adult attachment," *Journal of Personality and Social Psychology* 67, 430-445.

Gibson JJ (1979) *The ecological approach to visual perception*. Boston: Houghton Mifflin.

Giuliani A, Benigni R, Zbilut JP, Webber CL, Jr., Sirabella P, and Colosimo A (2002) Nonlinear signal analysis methods in the elucidation of protein sequence-structure relationships. *Chem. Rev.* 102 (5):1471-1492, 2002.

Glass L & Mackey MC (1988) *From clocks to chaos. The rhythms of life*. Princeton, N.J: Princeton University Press.

Goldberger AL (1992) Fractal mechanisms in the electrophysiology of the heart. *IEEE Eng Med.Biol.Mag.* 11 (2):47-52.

Goldberger AL, Amaral LA, Hausdorff JM, Ivanov PC, Peng CK, and Stanley HE (2002) Fractal dynamics in physiology: alterations with disease and aging. *Proc. Natl. Acad. Sci. U.S.A* 99 Suppl 1:2466-2472.

Goldberger AL, Peng CK, and Lipsitz LA (2002) What is physiologic complexity and how does it change with aging and disease? *Neurobiol.Aging* 23 (1):23-26.

Goldstein J (1994) *The unshackled organization. Facing the challenge of unpredictability through spontaneous reorganization*, Portland, Or: Productivity Press.

Gooddy W (1988) *Time and the nervous system*, New York: Praeger.

Groddeck GW (1923) *Das Buch vom Es. Psychoanalytische Briefe an eine Freundin*. Leipzig: Internationaler Psychoanalytischer Verlag

Guastello SJ (1995) *Chaos, catastrophe, and human affairs. Applications of nonlinear dynamics to work, organizations, and social evolution*, Mahwah, N.J: Lawrence Erlbaum Associates,.

Guastello SJ (2002) *Managing emergent phenomena. Nonlinear dynamics in work organizations*, Mahwah, N.J:Lawrence Erlbaum Associates, Publishers.

Have Pt (1999) *Doing conversation analysis. A practical guide*. London: Sage Publications.

Hazan C & Shaver PR (1987) "Romantic love conceptualized as an attachment process," *Journal of Personality and Social Psychology 52,* 511-524.

Heisenberg W (1958) *Physics and philosophy, the revolution in modern science*. (1st ed. ed.) New York: Harper.

Heisenberg W & Hayes F (1952) *Philosophic problems of nuclear science*. London: Faber and Faber.

Hendrick SS & Hendrick C (1995) "Gender differences and similarities in sex and love," *Personal Relationships* 2, 55-65.

Hennig R (1896) Entstehung und Bedeutung der Synopsien. *Zeitschrift für Psychologie*; vol. 10: 181- 222.

Hofer MA (1984) Relationships as regulators: a psychobiologic perspective on bereavement. *Psychosom.Med.* 46 (3):183-197.

Hofer MA (1994) Hidden regulators in attachment, separation, and loss. *Monogr Soc.Res.Child Dev.* 59 (2-3):192-207.

Hagege C (1993) *The language builder, an essay on the human signature in linguistic morphogenesis*. Amsterdam: J. Benjamins.

Hjelmslev L (1970) *Essais linguistiques*, Copenhague: Nordisk Sprogog Kulturforlag.

Hjelmslev L (1973) *Essais linguistiques II*, Copenhague: Nordisk Sprogog Kulturforlag.

Hoffmeyer J (1997) Biosemiotics: Towards a New Synthesis in Biology. *European Journal for Semiotic Studies* 9 (2).

Husserl E (1964) *The phenomenology of internal time-consciousness*, Bloomington: Indiana University Press.

Jacobson R (1966) *Essais de linguistique generale*, Paris F: Editions de Minuit.

James W (1920) *Collected essays and reviews*, London: Longmans, Green and Co.

James W (1967) *The writings of William James, a comprehensive edition*, New York: Random House.

Jones FJ (1995) *The Structure of Petrarch's Canzoniere,* Cambridge: Brewer.

Kachele H (2001) Are there "Pillars of Therapeutic Wisdom" for Psychoanalytic Therapy? *Journal of European Psychoanalysis* 12/13.

Kant I (1998) *Critique of pure reason*. Cambridge: Cambridge University Press.

Kant I (1987) *Critique of judgment*. Indianapolis, Ind: Hackett Pub. Co.

Kauffman SA (1993) *The origins of order. Self organization and selection in evolution*, New York: Oxford University Press.

Kauffman SA (1995).*At home in the universe. The search for laws of self-organization and complexity*, New York: Oxford University Press.

Khibnik AI, Kuznetsov YA, Levitin VV & Nikolaev EV (1993) "Continuation techniques and interactive software for bifurcation analysis of ODEs and iterated maps," *Physica D* 62, 360-37 1.

Kobak RR & Hazan C (1991) "Attachment in marriage: the effect of security and accuracy of working models," *Journal of Personality and Social Psychology* 60, 86 1-869.

Kosslyn SM (1980) *Image and mind*. Cambridge, Mass: Harvard University Press.

Kuhn TS (1996) *The structure of scientific revolutions*, Chicago, Il: University of Chicago Press,.

Kupfer DJ, Monk TH, Barchas JD, and MacArthur Foundation Network on the Psychobiology of Affective Disorders. (1988) *Biological rhythms and mental disorders*, New York: Guilford Press.

Lacan J (1966) *Écrits*, Paris:Editions du Seuil.

Lalande A & Société française, d. p. (1962). *Vocabulaire technique et critique de la philosophie*. (9 ed.) Paris: Presses Universitaires de France.

Laplanche J, Pontalis JB (1968) *Vocabulaire de la psychanalyse*. Paris: Presses Universitaires de France.

Leibniz GW (1996) *New essays on human understanding*. Cambridge: Cambridge University Press

Lévi-Strauss C (1958) *Anthropologie structurale*. Paris: Plon.

Lévi-Strauss C (1964) *Mythologiques*. Paris: Plon.

Levi-Strauss C (1964) Le Cru et le Cuit. Paris F: Librairie Plon.

Libet B, Freeman A, and Sutherland K (1999) *The volitional brain towards a neuroscience of free will*, Thorverton, UK: Imprint Academic,

Libet B (1993) *Neurophysiology of consciousness, selected papers and new essays*, Boston: Birkhauser.

Locke J (1694) *An essay concerning humane understanding*. (2d ed.) London: Printed for Awnsham and J. Churchil, at the Black Swan.

Luenberger DG (1979) *Introduction to Dynamic Systems,* New York: Wiley.

MacWhinney B & Bates E (1989) *The Crosslinguistic study of sentence processing*. Cambridge: Cambridge University Press.

Mahler MS (1968) *On human symbiosis and the vicissitudes of individuation*. New York: International Universities Press.

Mahler MS, Bergman A, & Pine F (1975) *The psychological birth of the human infant symbiosis and individuation*. New York: Basic Books.

Mainzer K (1994) *Thinking in complexity, the complex dynamics of matter, mind, and mankind*. Berlin: Springer-Verlag.

Margulis L (1981) *Symbiosis in cell evolution. Life and its environment on the early earth*, San Francisco: W. H. Freeman and Co.

Margulis L and Sagan D (1986) *Origins of sex three billion years of genetic recombination*, New Haven, Conn: Yale University Press.

Margulis L and Schwartz K V (1998) *Five kingdoms an illustrated guide to the phyla of life on earth*, New York: Freeman.

Marty P & de M'Uzan M (1978) Operative thinking (pensee operatoire). *Psyche (Stuttg), 32,* 974-984.

Matte Blanco I (1975) *The unconscious as infinite sets: an essay in bi-logic*, London: Duckworth.

Matte Blanco I (1988) *Thinking, feeling, and being. Clinical reflections on the fundamental antinomy of human beings and world*, London: Routledge.

Maturana HR & Varela FJ (1980) *Autopoiesis and cognition, the realization of the living*. Dordrecht, Holland: D. Reidel Pub. Co.

Mauss M (1968) *Œuvres*. Paris: Éditions de Minuit.

Merleau-Ponty M (1945) *Phénoménologie de la perception*. Paris: Gallimard.

Merleau-Ponty M (1962) *Phenomenology of perception*, New York: Humanities Press.

Merleau-Ponty M (1960) *La structure du comportement*, Paris: Presses Universitaires de France.

Merriam-Webster I (1998). *Merriam-Webster's collegiate dictionary*. (10th ed ed.) Springfield, Mass: Merriam-Webster.

Mill JS (1843) *A system of logic, ratiocinative and inductive, being a connected view of the principles of evidence, and methods of scientific investigation*. London: J. W. Parker.

Minsky ML (1968) *Semantic information processing*. Cambridge, Mass: MIT Press.

Minsky ML, Shore M, Marion A, & Voyager Company (1994) *First person, Marvin Minsky the society of mind* [Computer software]. New York: Voyager.

Morgan CL (1923) *Emergent evolution, the Gifford lectures, delivered in the University of St. Andrews in the year 1922*. London: Williams and Norgate.

Nagel E (1961) *The structure of science, problems in the logic of scientific explanation*. New York: Harcourt, Brace & World.

Nemiah JC (1977) Alexithymia. Theoretical considerations. *Psychother. Psychosom., 28,* 199-206.

Nemiah JC (1996) Alexithymia: present, past--and future? *Psychosom.Med., 58,* 217-218.

Nietzsche FW (2000) *The birth of tragedy*. Oxford: Oxford University Press.

Orsucci F (1981) Linguistic signs in neurosis and psychosis. *Il Lavoro Neuropsichiatrico* 68 [1/2].

Orsucci F, Giuliani A, Walter K, Webber C, and Zbilut J (1996) The prosodic-stylistic structure of human speech. *Int.J.Chaos Th.& A.* 2(3).

Orsucci F (1996a) Chaos and complexity in psychopathology. *Giorn.Ital.Psicopat.* 3 (2).

Orsucci F (1996b). Behcet's disease and psychosomatic patterns of thinking. A single case study. *Psychother.Psychosom.* 65 (2):112-114.

Orsucci F (Ed.)(1998) The complex matters of the mind, Singapore: World Scientific Pub.

Orsucci F (1999) Origins of the origins: evolution in semiotic universe. *Psicoloquy,* 99.10.013

Orsucci F (2000) Olfaction: New Frontiers for Cognitive Sciences. in: Sensors and Microsystems, Proceedings of the 5th Italian Conference — Extended to Mediterranean Countries. (edited by) D'Amico A & Siciliano P , Di Natale C. Singapore: World Scientific Publishing. .

Orsucci F (2002, in press) Linguistic landscapes. *Chaos & Complexity Letters* 1[2].

Ovidius PN (1977) *Ovid's Art of love, in three books, together with his Remedy of love part first*, New York: AMS Press,.

Palombo SR (1999) *The emergent ego. Complexity and co-evolution in the psychoanalytic process*. Madison, Conn: International Universities Press.

Pask G (1976) *Conversation theory. Applications in education and epistemology*. Amsterdam: Elsevier.

Paulesu E & et al (2000) A cultural effect on brain function. *Nat Neurosci*, Jan 3:91-96.

Pecora LM and Carroll TL (1990) Synchronization in chaotic systems. *Physical Review Letters* 64 (8):821-824.

Pecora LM and Carroll TL (1991). Driving systems with chaotic signals. *Physical Review A* 44 (4):2374-2383.

Peirce CS (1931) *Collected papers of Charles Sanders Peirce*. Cambridge: Harvard University Press.

Peirce CS (1953). *Letters to Lady Welby*. New Haven: Published by Whitlock for the Graduate Philosophy Club of Yale University.

Petitot J (1999) *Naturalizing phenomenology issues in contemporary phenomenology and cognitive science*, Stanford, Cal: Stanford University Press.

Pezard L, Nandrino JL, Renault B, el Massioui F, Allilaire JF, Muller J, Varela F, and Martinerie J (1996) Depression as a dynamical disease. *Biol.Psychiatry* 39 (12): 991-999.

Pezard L and Nandrino JL (2001) Dynamic paradigm in psychopathology: "chaos theory", from physics to psychiatry. *Encephale* 27 (3): 260-268.

Piaget J, Chomsky N, and Piattelli-Palmarini M (1980) *Language and learning the debate between Jean Piaget and Noam Chomsky*, Cambridge, Mass: Harvard University Press.

Pikovsky A, Rosenblum M, and Kurths J (2001). *Synchronization: a universal concept in nonlinear sciences*, Cambridge:Cambridge University Press.

Pinker S (1997) *How the mind works*, New York: WW Norton,.

Plato & Hermann KF (1921) *Platonis dialogi*. Lipsiae: in aedibus B. G. Teubneri.

Pylyshyn ZW (1984) *Computation and cognition: toward a foundation for cognitive science*. Cambridge, Mass: MIT Press.

Pyragas K (1996) Weak and strong synchronization of chaos. *Physical.Review E.* 54 (5):R4508-R4511.

Pyragas K (2001) Control of chaos via an unstable delayed feedback controller. *Phys.Rev.Lett.* 86 (11):2265-2268.

Quillian RM (1968) Semantic memory. in Minsky (Ed.) *Semantic Information Processing*, Cambridge: MIT Press.

Radzicki MJ (1993) "Dyadic processes, tempestuous relationships, and system dynamics," *System Dynamics Review 9,* 79-94.

Rensing L, an der Heiden U, Mackey MC (Eds) (2001) *Temporal Disorders in Human Oscillatory Systems.* Berlin: Springer.

Rinaldi S & Farina L (1995) *Positive Linear Systems: Theory and Applications*, Torino: Utet.

Rinaldi S (1996a) "Love dynamics: the case of linear couples," Working Paper WP-96-68, International Institute for Applied Systems Analysis, Laxenbug, Austria.

Rinaldi S (1996b) "Laura and Petrarch: an intriguing case of cyclical love dynamics," Working Paper WP-96-67, International Institute for Applied

Systems Analysis, Laxenbug,.

Rinaldi S & Gragnani A (1996) "Love dynamics between secure individuals: a modeling approach," Working Paper WP-96-69, International Institute for Applied Systems Analysis, Laxenbug.

Saussure Fd, Bally C, Sechehaye A, & Riedlinger A (1986) *Course in general linguistics*. LaSalle, Ill: Open Court.

Scharfe E & Bartholomew K (1994) "Reliability and stability of adult attachment patterns," *Personal Relationship* 1, 23-43.

Schiepeck G (1990) Synergetics in psychiatry. In: *Synergetics of cognition*, edited by H Haken and M Stadler, Berlin:Springer Verlag.

Schmid GB (1969) Chaos theory and schizophrenia: elementary aspects. *Psychopathology* 24 (4):185-198, 1991.

Schopenhauer A (1907) *The world as will and idea*. (6th ed ed.) London: K. Paul, Trench, Trübner & co., ltd.

Schrödinger E (1967) *What is life? & Mind and matter*. Cambridge: University Press.

Serres M & Authier M (1995) *A History of scientific thought, elements of a history of science*. Oxford: Blackwell

Searle JR, Dennett DC, and Chalmers DJ (1997) *The mystery of consciousness*, New York: New York Review of Books.

Searles HF (1979) *Countertransference and related subjects*. New York: International Universities Press.

Simpson JA, Gangestad SW & Lerma M (1990) "Perception of physical attractiveness: mechanism involved in the maintenance of romantic relationships," *Journal of Personality and Social Psychology 59,* 1192-1201.

Sonesson G (1989). *Pictorial concepts. Inquiries into the semiotic heritage and its relevance to the interpretation of the visual world*. Lund, Sweden: Lund University Press.

Spencer-Brown G (1994) *Laws of form*. (Limited ed ed.) Portland, Or: Cognizer Co.

Spitz RA (1969) Aggression and adaptation. *J.Nerv.Ment.Dis.* 149 (2):81-90.

Sprecher S, Quintin S & Hatfield E (1994) "Mate selection preferences: gender differences examined in a national sample," *Journal of Personality and Social Psychology* 66, 1074-1080.

Stein AH (1998) Epistemological and treatment implications of nonlinear dynamics. In Orsucci F (ed) *The complex matters of the mind*. Singapore: World Scientific.

Strogatz SH (1988) Love affairs and differential equations, *Mathematics Magazine* 61, 35.

Strogatz SH (1994) *Nonlinear Dynamics and Chaos with Applications to Physics, Biology, Chemistry and Engineering* (Addison-Wesley, Reading, MA).

Thomasson N and Pezard L (1999) Dynamical systems and depression: a framework for theoretical perspectives. *Acta Biotheor.* 47 (3-4):209-218.

Thompson E and Varela FJ (2001) Radical embodiment: neural dynamics and consciousness. *Trends Cogn Sci.* 5 (10):418-425.

Tustin F (1990). *The protective shell in children and adults*. London: Karnac.

Varela FJ, Thompson E, and Rosch E (1991) *The embodied mind cognitive science and human experience* , Cambridge, Mass: MIT Press.

Varela FJ (1999) Cognition without representations. *Riv.Biol.* 92 (3):511-512.

Varela FJ, Lachaux JP, Rodriguez E, and Martinerie J (2001) The brainweb: phase synchronization and large-scale integration. *Nat.Rev.Neurosci.* 2 (4):229-239.

Verhulst F (1994) Metaphors for psychoanalysis. *Nonlinear Science Today* 4 (1):16.

Vitiello G (2001) *My double unveiled, the dissipative quantum model of brain*, Amsterdam: John Benjamins Pub. Co.

West BJ, Zhang R, Sanders AW, Miniyar S, Zuckerman JH, an Levine BD (1999) Fractal fluctuations in cardiac time series. *Physica A* 270 (3-4):552-566.

West BJ (1985) *An essay on the importance of being nonlinear*, Berlin: Springer Verlag.

West BJ (1993) *Patterns, information and chaos in neuronal systems*, Singapore: World Scientific.

West BJ and Deering WD (1995) *The lure of modern science, fractal thinking*, Singapore: World Scientific Pub.

West BJ (2001) Fractal probability measures of learning. *Methods* 24 (4):395-402.

West ML (1983) *The Orphic poems*. Oxford: Oxford University Press.

Winnicott DW (1971a) *Playing and reality*, London: Tavistock Publications.

Winnicott DW (1971b) *Therapeutic consultations in child psychiatry*, London: Hogarth Press for the Institute of Psycho-Analysis.

Wittgenstein L (1961) *Tractatus logico-philosophicus*. London: Routledge & Kegan Paul.

Wittgenstein L (1967) *Philosophical investigations*. Oxford.

Wittgenstein L & Rhees R (1983) *Remarks on Frazer's Golden bough*. Retford, Nottinghamshire, England: Brynmill Press.

Zbilut JP, Sirabella P, Giuliani A, Manetti C, Colosimo A, and Webber CL Jr (2002) Review of nonlinear analysis of proteins through recurrence quantification. *Cell Biochem.Biophys.* 36 (1):67-87.

Index